Creativity and Culture in Greater China

The Role of Government, Individuals, and Groups

Edited by

Chi-Cheung Leung
Sonny Shiu-Hing Lo

Los Angeles

Creativity and Culture in Greater China: The Role of Government, Individuals, and Groups
Copyright © 2014
By Chi-Cheung Leung, Sonny Shiu-Hing Lo

Distributed by Transaction Publishers
10 Corporate Place South, Suite 102
Piscataway, NJ 08854

All rights reserved. Exclusive English language rights are licensed to Bridge21 Publications, LLC. No part of this book may be used or reproduced in any matter whatsoever without written permission from the publisher except in the case of brief quotations embodied in critical articles and reviews.

For information contact Bridge21 Publications, LLC, 11111 Santa Monica Blvd, Suite 220, Los Angeles, CA 90025.

Published in the United States
Cover Design by Chi-Wai Li
Interior Layout by Mehboob Sam
Copyedited by Peg Goldstein
ISBN 978-1-62643-008-2 Paperback

Table of Contents

Acknowledgements 15

Abbreviations 17

Contributors 19

Introduction

Toward an Analytical Framework of Understanding the Development of Creative Industries: A Tripod of the Government, Individuals, and Groups 23
    Sonny Shiu-Hing Lo and Chi Cheung Leung

Section 1: China

Chapter 1

Scrambling for the Mainland China Market 47
    Michael Keane

Chapter 2

Policy Transfer of Creative Industries for Economic Development: The Chinese Metropolitan Experience 65
    Xuan-Olivia Jiang

Chapter 3

Cultural Revitalization and Pedestrianization in Chengdu and Kunming: Lanes and Streets 85
    Anthony K. C. Ip

Chapter 4

Artistic Creativity in Modern Art Education: The Case of the Shanghai Art College, 1913–1937 115
    Jane Zheng

Chapter 5

Combating Cultural Heritage Crimes: Recent Developments in China 133
    Minxing Zhao

Section 2: Hong Kong

Chapter 6

The Role of Hong Kong Government in Creative Industries: Analyzing the Cultural Discourse of Three Policy Addresses .. 147
    Ming Lai and Chi Cheung Leung

Chapter 7

Challenges of Museums in Hong Kong in Promoting Local Cultural Heritage: An Assessment of Hong Kong International Museum Day (2001–2010) 161
    Wai Shing Lee

Chapter 8

Revival of Chinese Calligraphy in Hong Kong's Architecture .. 183
    Roger Tin Sing Kho

Chapter 9

Venturing Strategies in Developing the Hong Kong Art Industry 201
    Yan Tung

Chapter 10

The Development of Cultural Entrepreneurship: Case Studies of Four Community Orchestras in Hong Kong 213
    Chi Cheung Leung

Chapter 11

The Birth and Development of a Cultural Advocacy Group in Hong Kong: Centre for Community Cultural Development.... 241
    Chiu Yu Mok and Eric Ng

Chapter 12

Culture and the Media
in a Global Age:
The Role of the Media in Shaping Cultural Development in
"Global City Wannabes" 253
    Vivienne Chow

Chapter 13

The Development of Creativity in Hong Kong: Technology, Talent, and Tolerance 267
    Victor Kwong

Section 3: Comparative and Cross-Border Perspectives

Chapter 14

Transparent Cosmopolitanism: New Versions of the
Flâneur in Taiwanese and Hong Kong Cinema 279
    Joey Moon

Chapter 15

Educating Entrepreneurial Portfolio Musicians for
Twenty-First-Century Careers 297
    Patrick M. Jones

Chapter 16

Inclusive Cultural Policy in the Music Education of Nanjing
and Hong Kong 313
    Rita Lai Chi Yip and Ji Hong Ye

Chapter 17

Reshaping the Creative Culture of Hong Kong Cinema under Chinese Shadows: Marketing Coproduction Films in Mainland China and the Preservation of Localism 327
    Patrick Yeuk-Kwong Yuen

Chapter 18

Political Comics and Freedom of Expression: The 2012 Chief Executive Election in Hong Kong and Implications for Mainland China.. 351
    Sonny Shiu-Hing Lo

Conclusion 369

Index 371

List of Tables

Table 0.1 Relations between creative and cultural industries and the economies of Greater China 31

Table 0.2 China's cultural and related industries classifications in 2012... 34

Table 1.1 The cultural innovation timeline 52

Table 5.1. Countries entered into bilateral agreements with China (in chronological order).. 138

Table 6.1 Frequency of appearance of culture-related keywords in policy addresses of chief executives in past years 148

Table 7.1 Numbers of participating units (2001–2010) 165

Table 7.2 Numbers of non-LCSD units (2001–2010) 165

Table 7.3 Numbers of visitors to seven museums (2002–2011) 174

Table 9.1 Application of venturing strategies for two private and two public cultural enterprises... 209

Table 10.1 Features of four community Chinese orchestras in Hong Kong 231

Table 10.2 Features of four community orchestra models 233

Table 13.1 Comparison of the creative class population and the whole working population.. 270

Table 13.2 Global technology rankings in Singapore and Hong Kong 272

Table 15.1 Undergraduate learning/ teaching recommendations for career development in music 302

Table 16.1 Qualitative research method in analyzing the curriculum.. 315

Table 17.1: Box office revenue of coproduced films in mainland China, 2005–2010 329

Table 17.2 Box office revenue of mainland China, 2005–2012. 331

Table 17.3 Top ten in box office revenue, 2012—domestic productions 332

List of Figures

Figure 0.1 An Analytical framework in the development of creative industries.. 25

Figure 2.1 The anatomized process of policy transfer 70

Figure 3.1 Wide and Narrow Lanes in the past 86

Figures 3.2.1 & 3.2.2 Wide and Narrow Lanes today................. 88

Figure 3.3 Integrating old and new 93

Figures 3.4 & 3.5 Cultural motif and articulation of architectural and landscape spaces 94

Figure 3.6 Chengdu: Wide and Narrow Lanes urban design and planning concept.. 95

Figure 3.7 Kunming old city wall 98

Figure 3.8 Former Young Men's Christian Association 100

Figure 3.9 Old Street pedestrianization....................................... 103

Figures 3.10 & 3.11 Activities along the pedestrian zone 106

Figure 3.12 Zhongai Gate (Righteous and Love)....................... 106

Figure 3.13 Kunming: Old Street urban design and planning concept.. 107

Figure 7.1 Awareness level and participation rate of education and public programmes organized by the museums................. 170

Figure 7.2 Factors affecting the decision of visit to the museums... 172

Figure 8.1 Mourning letter on the assassination of Yang Quyun by Sun Yatsen at Pak Tze Lane Park. Photograph by Roger T. S. Kho. 187

Figure 8.2 A giant rock with Sun Yatsen's calligraphy, reading, "The world is for all" (天下為公, tian xia wei gong). Photograph by Roger T. S. Kho. 188

Figure 8.3 The stone gateway at Sun Yatsen Memorial Park with regular script calligraphy. Photograph by Roger T. S. Kho 188

Figure 8.4 The stone gateway with seal script calligraphy at Hollywood Road Park. Photograph by Roger T. S. Kho. 190

Figure 8.5 The standard script stone tablet at Hollywood Road Park. Photograph by Roger T. S. Kho. 190

Figure 8.6 Two carved granite plaques unearthed from the original South Gate in Kowloon Walled City Park. Photograph by Roger T. S. Kho. 191

Figure 8.7 Fist calligraphy (拳書) of the character shou (壽) by Zhang Yutang (張玉堂). Photograph by Roger T. S. Kho ... 191

Figure 8.8 Fist calligraphy (拳書) of the character moyuan (墨緣) by Zhang Yutang (張玉堂). Photograph by Roger T. S. Kho.... 192

Figure 8.9 Narrative on the North Gate of Kowloon Walled City Park by architect Tse Shunkai (謝順佳). Photograph by Roger T. S. Kho. 192

Figure 8.10 The Wisdom Path (心經簡林, 2005), with Heart Sutra calligraphy by Jao Tsung-I (饒宗頤). Photograph by Roger T. S. Kho. 193

Figure 8.11 Detail of Heart Sutra calligraphy. Photograph by Roger T. S. Kho. 194

Figure 8.12 "Goose" in one stroke (一字鵝) in Hau Wong Temple, Kowloon City. Photograph by Roger T. S. Kho 195

Figure 8.13 Clan association for the Fu family. Calligraphy written by Yu Youren (于右任). Photograph by Roger T. S. Kho. 197

Figure 8.14 Yee Hing Loong (義興隆), shop trading in antique classic furniture and Chinese arts and crafts. Calligraphy by Luo Shuzhong (羅叔重). Photograph by Roger T. S. Kho. 197

Figure 8.15 Coffin shop. Calligraphy by Ou Jiangku (區建公). Photograph by Roger T. S. Kho. 198

Figure 8.16 Building facade with works of calligraphy. Photograph by Roger T. S. Kho. 198

Figure 10.1 Development of cultural entrepreneurship 217

Figure 12.1 The Media & Cultural Ecology................................ 262

Figure 15.1 Number of roles for music-related income 299

Figure 15.2 Aggregated revenue streams of conservatory of music school graduates 300

Figure 16.1 Comparison of non-Chinese music materials for teaching in primary schools... 319

Figure 16.2 Non-Chinese classic or folk music enacted for each level of primary class (Nanjing).. 320

Figure 16.3 Non-Chinese classic or folk music enacted for each level of primary class (Hong Kong) 321

Figure 18.1 Zunzi, "Playing to the extent of serious injury" (尊子，玩殘), retrieved from http://www.civilhrfront.org/gallery/photos/zz10fan1p_b.jpg.. 352

Figure 18.2 Zunzi, "Each having a duty" (尊子，各司其職), retrieved from http://www.civilhrfront.org/gallery/photos/zz15fan1p_b.jpg.. 353

Figure 18.3 Zunzi, "Mouth Performance" (尊子，嘴皮表演), retrieved from http://www.civilhrfront.org/gallery/photos/zz17fan1p_b.jpg.. 353

Figure 18.4 Xiao Yu, "Cake-throwing warrior Russel Gordon" (小如，扔蛋糕鬥士盧爾　戈丹), retrieved from http://hk.epochtimes.com/archive/Issue76/szlz-1.html...................... 354

Figure 18.5 Zunzi, 尊子, retrieved from http://news.mingpao.com/20120516/gza1_image1.htm?Mode=1 354

Figure 18.6 Zunzi, 尊子, retrieved from http://www.rthk.org.hk/special/hkconnection/hksar10/guest17.htm 354

Figure 18.7 尊子，財爺加碼為六四降溫, retrieved from http://www.myradio.com.hk/talks/viewthread.php?tid=766586&extra=&page=23 355

Figure 18.8 Malone, 馬龍, retrieved from http://www.hkreporter.com/talks/thread-766586-68-1.html 355

Figure 18.9 Zunzi, 尊子，整色整水, retrieved from http://71peoplepile.blogspirit.com/archive/2005/04/21/424%E3%80%8C%E5%8F%8D%E5%B0%8D%E9%87%8B%E6%B3%95%E3%80%8D%E5%A4%A7%E9%81%8A%E8%A1%8C.html 356

Figure 18.10 Netizen, 網民，網民創意唐英年惡搞圖全集，Sina, retrieved from http://news.sina.com.hk/news/1603/3/2/2580142/1.html 357

Figure 18.11 Cuson Lo, album of 快樂政治, retrieved from https://www.facebook.com/#!/media/set/?set=a.10150370047404148.432130.657699147&type=3 357

Figure 18.12 Cuson Lo, album of 快樂政治, retrieved from https://www.facebook.com/#!/media/set/?set=a.10150370047404148.432130.657699147&type=3 357

Figure 18.13 Cuson Lo, album of 快樂政治, retrieved from https://www.facebook.com/#!/media/set/?set=a.10150370047404148.432130.657699147&type=3 358

Figure 18.14 Cuson Lo, album of 快樂政治, retrieved from https://www.facebook.com/#!/photo.php?fbid=10150819015104148&set=a.10150370047404148.432130.657699147&type=3&theater 358

Figure 18.15 Netizen, 網民，無待堂, retrieved from http://dadazim.com/journal/2012/03/dont-be-sad 359

Figure 18.16 Netizen, "According to March 21st 2012 Apple Daily," Hong Kong Internet News, retrieved from http://badcanto.wordpress.com/2012/03/25/the-death-of-hong-kong/ 359

Figure 18.17 Netizen, 網民，E-Zone, retrieved from http://www.e-zone.com.hk/discuz/viewthread.php?tid=56602 360

Figure 18.18 Cuson Lo, album of 快樂政治, retrieved from https://www.facebook.com/#!/media/set/?set=a.10150370047404148.432130.657699147&type=3 360

Figure 18.19 Netizen, 網民, "The Whole City Wanted, Henry Tang—Unauthorised building works. Breaking the law knowingly, Donald Tsang—Power abuse. Advantages transfer, CY Leung—Cover benefit up. False statement," Hong Kong Instant News, retrieved from http://badcanto.wordpress.com/2012/03/01/donald-tsang-cried-in-his-misconduct-hearing/ 361

Figure 18.20 Netizen, 網民, Oriental Daily News, retrieved from http://orientaldaily.on.cc/cnt/news/20111005/00176_005.html 361

Figure 18.21 Netizen, 網民，唐英年改圖大收集：電影海報篇, Internet Digest, retrieved from http://obelia2.blogspot.com.au/2012/02/blog-post_6346.html 361

Figure 18.22 Unknown author, C. Y. Leung to Ma Ying-jeou: "Join one country two systems. You are guaranteed to be elected!" Hong Kong Internet News, retrieved from http://badcanto.wordpress.com/2012/03/25/the-death-of-hong-kong/ 362

Figure 18.23 Unknown author, "Comrade Leung," Hong Kong Internet News, retrieved from http://badcanto.wordpress.com/2012/03/25/the-death-of-hong-kong/ 362

Figure 18.24 Unknown author, "Against me? You guys don't act foolishly!" Hong Kong Internet News, retrieved from http://badcanto.wordpress.com/2012/03/25/the-death-of-hong-kong/ 362

Figure 18.25 Zunzi, 尊子，retrieved from http://peepsen.blogspot.com/2012/04/blog-post_8532.html 363

Figure 18.26 Cuson Lo, album of 快樂政治, retrieved from https://www.facebook.com/#!/media/set/?set=a.10150370047404148.432130.657699147&type=3.... 363

Last accessed August 31, 2014

Acknowledgements

This book could not have been completed without the help of many scholars and friends. First and foremost, we express our gratitude to Professor Joshua Ka-Ho Mok for his support of our workshops on creativity in the Greater China. Two workshops were held at the Hong Kong Institute of Education in June 2012 and March 2013: "Creativity, Culture and Related Industries: Implication for Greater China Region" and "Transformation, Development and Culture in Asia: Multidisciplinary Perspectives." Secondly, we thank Bridge 21 for its support of our book project, especially Gregory, who provided constant advice and suggestions. Thirdly, we thank our research assistant Yan Tung, who helped us compile the manuscript in a professional way and whose networks were so critical to the last stage of the book project. Credit must be given to her indeed. Fourthly, we thank all the contributing authors for their work and insights, without which we could not have produced this meaningful work. Finally, we thank all those who have given us permission to reprint the photos, articles (the chapter on cultural entrepreneurship by Chi Cheung Leung was published in *Asian Education and Development* in December 2012), and political comics. We as the editors are responsible for the errors, but we do hope that this book is the first attempt at analyzing and studying the development of creative and cultural industries in Greater China (defined as mainland China, Taiwan, Hong Kong, and Macao) in a comprehensive and an innovative manner. As humble academics and editors, we expect this work to stimulate more research in the study of creativity and of creative and cultural industries in Greater China in the coming years, especially using the analytical framework that we advance in the introductory chapter.

Chi Cheung Leung and Sonny Shiu-Hing Lo
December 10, 2013

Abbreviations

Centre for Community Cultural Development (CCCD)
Chengdu International, Science, Education and Arts Town (CISEAT)
China Central Television (CCTV)
Chinese Communist Party (CCP)
Closer Economic Partnership Agreement (CEPA)
Creative Industries Mapping Document (CIMD)
Department for Culture, Media and Sport (DCMS)
foreign direct investment (FDI)
Fung Ying Seen Koon (FYSK)
Guangzhou Association for Social Sciences (GASS)
Hong Kong International Museum Day (HKIMD)
Hong Kong Special Administrative Region (HKSAR)
Hong Kong Taoist Orchestra (HKTO)
information and communication technology (ICT)
International Council of Museums (ICOM)
Leisure and Cultural Services Development (LCSD)
Lok Sum Chinese Orchestra (LSCO)
Macao Special Administrative Region (MSAR)
national innovation systems (NIS) or regional innovation systems (RIS)
National People's Congress (NPC) of China
New Tune Chinese Orchestra (NTCO)
own-brand manufacturing (OBM)
People's Republic of China (PRC)
Republic of China (ROC)
State Administration of Radio, Film and Television (SARFT)
United Nations Conference on Trade and Development (UNCTAD)
United Nations Educational, Scientific and Cultural Organization (UNESCO)
Yao Yueh Chinese Music Association (YYCMA)

Contributors

Vivienne Chow is a cultural journalist and critic and was named one of the world's best young journalists while representing Hong Kong at the 2004 inaugural Berlinale Talent Press at the Berlin International Film Festival. She has researched on post-colonial Hong Kong cultural policy for her master degree at the University of Hong Kong and is the founding director of non-profit educational initiative Cultural Journalism Campus. She is currently senior reporter covering cultural affairs at the South China Morning Post and publishes a blog Culture Shock at www.viviennechow.com.

Anthony K. C. Ip is vice president of the New Asia Arts and Business College and former program coordinator at the United International College (BNU-HKBU) in China's cultural industries management. He is a graduate of Massachusetts Institute of Technology and Louisiana State University and has more than thirty years of professional experience in Mainland China, Hong Kong, and North America.

Xuan-Olivia Jiang is an assistant professor at the Sun Yat-Sen University. She earned her PhD from the University of Delaware, with a dissertation about policy diffusion in creative industries. Her current research interests are industrial policy, the relationship between government and enterprises, and the relationship between government and the market.

Patrick M. Jones is professor and director of Syracuse University's Setnor School of Music. He has published articles on a variety of topics, including music education history, curriculum, policy, and theory. He is currently chair of the Policy Commission of the International Society for Music Education.

Michael Keane is principal research fellow at the Australian Research Council Centre of Excellence for Creative Industries and Innovation (CCI) at the Queensland University of Technology, Brisbane.

Roger Tin Sing Kho is an MPhil candidate at the Academy of Visual Arts at Hong Kong Baptist University. His current research is on the identity creation of Chinese calligraphy works in Hong Kong's urban area. With a first degree in architecture and a particular interest in Chinese calligraphy, he focuses his research on identity creation through public artworks in urban areas.

Victor Kwong is an executive officer at the Hong Kong Institute of Education. He obtained his MA in sociology from the Chinese University of Hong Kong and his BSocSc in international studies from the City University of Hong Kong. His research interests focus on institutional research, strategic planning in higher education institutions, and the sociology of education.

Ming Lai is a postdoctoral fellow at the Centre for Learning, Teaching and Technology at the Hong Kong Institute of Education. He is an amateur Cantonese lyricist; his Cantonese version of "Just the Way You Are" (Bruno Mars), sung by Robynn Yip, has been viewed over sixty thousand times on YouTube.

Wai Shing Lee is currently an assistant instructor at the Lingnan Institute of Further Education in Hong Kong. He has a range of research interests, including the history of Hong Kong, the history of Sino–Japanese relationships, and the museum sector in Hong Kong. He is now undertaking a project on the relationship between museum and identity-building in Hong Kong.

Chi Cheung Leung, a prolific composer and maestro conductor, is associate dean of the Faculty of Liberal Arts and Social Sciences and associate professor in the Cultural and Creative Arts Department at the Hong Kong Institute of Education. He is an expert in cultural entrepreneurship, Chinese music, and music education. His recent books include *A Study on the Music of Shijing and Chuci* and *Music Education Policy and Implementation: International Perspective*.

Sonny Shiu-Hing Lo is professor and head of the Department of Social Sciences at the Hong Kong Institute of Education. He is an expert in Hong Kong and Macao politics, cross-border crime in Greater China, and policing in China and Hong Kong. His recent books include *The Politics of Cross-Border Crime in Greater China* and *The Dynamics of Beijing-Hong Kong Relations*. His new single-authored book will focus on the politics of China's earthquake management.

Joey Moon, also known as Tzong-Huei Her, obtained her BA in Spanish literature and language from Providence University in Taiwan and an MA in international business administration from Bournemouth University in the United Kingdom. She is now a master's-degree candidate in cinema studies in the Department of Motion Pictures at the National Taiwan University of Arts. Joey has been deeply influenced by the works of Dickens, Austen, Kurosawa, and Truffaut. Her lifetime goal is to write literary and film critiques.

Chiu Yu Mok is a founder of the Centre for Community Cultural Development and the Asian People's Theatre Festival Society. Presently he is chief executive of the centre and chairperson of the festival society. He is a people's theater worker, a performance artist, and a community cultural development worker. He has organized many cross-cultural theater collaborations and was codirector of the International Drama/Theatre and Education Association Congress in 2007. He received the Drama Achievement Award from the Hong Kong Arts Development Council in 1999.

Eric Man Kei Ng was a project officer and project manager of the Centre for Community Cultural Development. He was awarded the Clore Scholarship for the 2013–2014 year. He is an actor and an experienced arts administrator, having organized many international collaborations, including the Confluence of Rivers, which that toured Hong Kong, Thailand, Bangladesh, Uganda, England, and France.

Yan Tung has worked in number of arts organizations in Hong Kong, China, and Australia, holding roles in curatorial, educational, and marketing areas for the last ten years. She holds master's degrees in arts administration and fine art and is currently a researcher with the Faculty of Cultural and Creative Arts at the Hong Kong Institute of Education.

Jihong Ye is a distinguished professor and head of the Music Education Department at the Nanjing Institute of Arts. She is also a member of the Music Education Committee and Vocal Music Committee of the Education Society in China. A graduate of the Berlin University of Arts, she was a visiting scholar in Holland. She has obtained special funding from the Chinese Ministry of Education, worked collaboratively with German and Chinese music institutions, and secured a scholarship for leading artists from the Austrian federal government.

Rita Lai Chi Yip is an assistant professor at the Hong Kong Institute of Education. She is a coauthor of Chinese music CD-ROMs and an editor of books. She has published numerous papers and is a commissioner of the Music Policy Commission of the International Society of Music Education. She has also assisted the Curriculum Development Council and the Hong Kong Examinations and Assessment Authority in formulating music curriculum guides for schools.

Patrick Yeuk-Kwong Yuen has a strong interest in media studies, focusing on creative culture and industry. He studied film in the United States and received his master's degree in cinema studies from New

York University. After years of experience in film and TV production and journalism, he currently teaches in Macao while engaging in creative writing.

Minxing Zhao is a PhD candidate in the Department of Social Science at the Hong Kong Institute of Education. He earned his master's degree in Political Science at the University of Waterloo and a law degree in mainland China. His research interests include comparative politics, social and political change in Greater China, transnational crimes, and international relations in Northeast Asia.

Jane Zheng is an assistant professor in the Cultural Management Program at Chinese University of Hong Kong. She received her master's degree from the Fine Arts Department and her doctoral degree from the Architecture Department at the University of Hong Kong. Her ongoing interests include planning for creative cities, cultural development and policies in China, and twentieth-century Chinese painting history.

Introduction

Toward an Analytical Framework of Understanding the Development of Creative Industries: A Tripod of the Government, Individuals, and Groups

Sonny Shiu-Hing Lo and Chi Cheung Leung

This publication is a collection of thoughts from scholars and researchers on how to describe and critically evaluate the current development of creative industries in the Greater China Region. Stemming from the conferences "Creativity, Culture and Related Industries: Implication for Greater China Region" and "Transformation, Development and Culture in Asia: Multidisciplinary Perspectives," held at the Hong Kong Institute of Education in June 2012 and May 2013, the revised book chapters address creative and cultural industries and shed light on content production, distribution, public perception, and relationships between sectors that are seemingly separate but could be reconnected through policy changes. Writers from a number of creative fields have contributed valuable insights with case studies and figures to show the phenomena, utilizing arguments in different disciplines to facilitate regional understanding of creative industries, with the hope that market or government intervention in identified areas can minimize hindrance to creativity growth and can address the weak areas.

From historical references to the commercialization of cultural products, the chapters adopt a multidisciplinary, cross-cultural, and experimental approach to dissect the industry into a number of different art forms, including art and music, popular culture, psychology, entrepreneurship, economic sectors, and political expressions in the changing social and political context of Greater China, an area that is still under development and in need of systematic research. Greater China in this book refers to the People's Republic of China (PRC), the Republic of China (ROC) on Taiwan, Hong Kong, and Macao. The various arts forms in this region, including their promotion, preservation, rejuvenation, and policy transfers, are addressed in this book. The importance of audience participation interwoven throughout the planning for all sorts of arts and cultural programs is emphasized, including the survival of smaller art organizations that address market needs. Cultivating cultural values among the wider public can sustain the appreciation of cultural heritage, especially through the means of adopting technology. There are reflections and introspections on creativity and cultural management issues, including art education practices and curriculum reforms.

CREATIVITY AND AN ANALYTICAL FRAMEWORK IN THE DEVELOPMENT OF CREATIVE INDUSTRIES

Creativity can be defined as ideas and practices that are both new and innovative, contributing to the development of the cultural, economic, social, technological, and political dimensions of a city, nation, or state. The human mind is the source of creativity, the applications of which sustain the continual growth of mankind. Creativity can be applied to different areas, professions, and disciplines, including creative and cultural industries. If creativity refers to new and innovative ideas put into practice, then the scope of creative industries can be very broad, ranging from arts to music, from cultural heritage to cultural policy, from eco-tourism to technological advancement, from societal expressions to freedom of thought, and from political comics to political reforms.

To understand the development of creative industries, we advance a tripod involving the roles of government, individuals, and groups. Their triangular relations determine the ways in which creative industries can and will be developed and the extent to which creativity can be generated and enhanced.

Governments are, by definition and in practice, more bureaucratic than creative, because creativity can entail critical thinking that challenges traditional ways of doing things, including governance and policies. Governments are forced to be creative when they are confronted with technological challenges that call for new ideas, practices, and policies. For example, the emergence of e-government can be seen as a response to the globalization of technological advancement. Government departments have to create their own websites in response to the needs and demands of the populace, thus narrowing the gap between the rulers and the ruled.

However, not all governments are creative, because ideologies can shape the extent of creativity. Capitalist states that attach importance to the role of market forces tend to see creativity as a means to enhance economic competitiveness and social diversity. However, socialist and communist states, which attach greater importance to public power or to public ownership of the means of production, may see creativity as potentially politically incorrect or dangerous. Indeed, the real world of democracy, as C. B. Macpherson had long stated, includes capitalist states, socialist and communist states, and developing states.[1] He said the real world of democracy maximizes the potential of individuals, implying that creativity is actually the key to success in every type of democracy, including socialist and communist, capitalist, and developing ones. Arguably, creativity in governance is an ideal that remains to be fully realized in every society and polity.

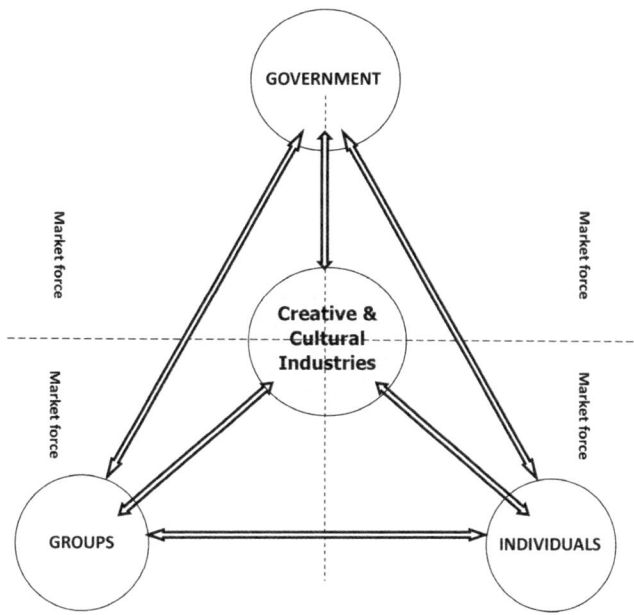

Figure 0.1 An Analytical framework in the development of creative industries

The role of government in the development of creative industries is complex, apart from the influential factor of state ideology as mentioned above. First, government may formulate and implement policies in favor of the growth of creative industrial sectors, ranging from cultural heritage to arts, from paintings to antiques, from music to comics, from science parks to museums, and from sports to historical sites for tourists. Second, government can use subsidies to stimulate the growth of creative industries. Third, government can enact laws and regulations to protect cultural heritage and historical sites. Fourth, government may intervene in shaping the market of a creative industry so that more audience participation can be encouraged and mobilized. Fifth, government may create statutory or public bodies and organizations solely responsible for the promotion of a particular cultural sector, such as filmmaking or performing arts, for the sake of engendering public support for the targeted creative industry. Sixth, government may make use of existing or new educational institutions to promote a particular cultural and creative industry through infrastructure and project investments. In a nutshell, government plays multidimensional roles in facilitating the metamorphosis of creative industries through financial means, policy tools, legal instruments, market intervention, organizational revamping, and educational measures.

The role of individuals and groups cannot be ignored in the development of creativity. Individuals indeed have the intelligence, wisdom, and power to shape society. Individual contributions to creativity are enormous, including the design of new ideas and ways of improving and perfecting products, art forms, music, technology, social values, and political systems. The creative power of individuals has never been explored fully. Talents in the diverse fields of creativity contribute immensely to the development of each industry. With the rapid advancement of technology and better educational opportunities in developed and developing countries, individual creativity continues to excel. However, the prosperity of a particular creative industry does not necessarily reflect an increase in the income of individual talents. In most cases, they receive unstable income and depend heavily on freelance work. On the other hand, the increasing popularity of Internet technology has generated a major channel for these talents to demonstrate and showcase their creativity through social media. Many of their creative products outshine traditional products and can even produce new markets, the impacts of which cannot be underestimated. Individuals can be their own "solopreneurs." Yet individual ideas that contribute to creativity are premised on the fact that freedom of speech, freedom of expression, and freedom of thought persist. To the extent that these various forms of freedom exist and thrive, individual creativity can be maximized further.

Similarly, group work can contribute immensely to the development of creativity. Groups are formed by individuals with common objectives and shared interests. These groups can take the form of an interest group, an orchestra, or even a private museum. Groups are composed of individuals whose potentials can be maximized collectively. Groups can also be referred to as entrepreneurs, depending on how individuals in the group manage their business. Recent studies have highlighted the importance of entrepreneurship in developing cultural and creative industries. The revenue generated from these industries continues to grow in some countries, sometimes despite fluctuations in the global economy. Even with a slight decrease, the economic growth of creative industries is able to pick up fast. Thus creative industries have emerged as an attraction for investment. With appropriate investment strategies and management skills, entrepreneurs, artists, and various stakeholders can collaborate to explore and invent their markets. The existence of the markets provides spaces for government, individuals, and groups to develop creative and cultural industries.

It is noteworthy that groups and individuals may be either professionals or amateurs in creative industries. While professionals use their professionalism and dedication in promoting cultural industries, amateurs may seek various means of promoting their creative and cultural industries in innovative ways beyond the expectations and support of the government. The lack of government support may even provide an impetus for individuals and groups to develop their own creative industries. Independence from the government, both financially and in terms of skills, may provide more social and political space for them to express ideas autonomously. The abundant cultural resources available in the community provide ample opportunities for collaboration and interaction for the development of creative and cultural industries.

Finally, the market plays a crucial role in filling the gap between the government on the one hand and individuals and groups on the other. In the event that government support of a creative industry is lacking, this gap can be filled by the market as long as it produces the audience, participants, and consumers supportive of a young but growing creative industry. In other words, the market itself is a powerful and unpredictable factor shaping the popularity of particular creative industries. Even if the government is by no means enthusiastic in financially supporting a creative sector, the market alone can propel the development of the sector and transform it into a highly popular and successful creative industry. The "invisible hand" of the market can compensate for the absence of government support for a certain creative industry, and it can have profound impacts on the sustainability of the momentum for developing the creative sector concerned. Simultaneously, government, individuals, and groups can be proactive, interactive, and reactive in designing strategic ways of overcoming barriers and initiating ventures to respond to the needs of markets.

Chapters in this book focus on the varying roles of government, individuals, and groups in mainland China, Taiwan, Hong Kong, and Macao. Section 1 is devoted to the study of creative industries in the PRC. Section 2 analyzes examples from Hong Kong. Section 3 is comparative and cross-boundary, exploring different cities that have developed cultural and creative industries.

From a broadly comparative perspective, the PRC is a socialist state that attaches importance to state intervention in the market, political correctness, and the economic utility of creative industries. The socialist government is interventionist in shaping the role and popularity of creative industries. Creativity, politically speaking, has to serve the ruling regime and party in power, namely the Chinese Communist Party (CCP). Hence the development of creative industries in the PRC naturally has political and economic objectives: supporting the policies of the CCP in achieving not only economic modernization but also China's rise, and facilitating the PRC's economic transition from communism under the Mao Zedong years to socialist marketization since the Deng Xiaoping era. Apparently, the development of creative industries in China is highly economic, but actually it is hidden political. However, given the rapid economic and social development in China, the growth of creative industries will face tremendous changes.

In the Hong Kong Special Administrative Region (HKSAR), the government has traditionally adopted a policy of positive noninterventionism in the economic market. Similarly, this ideology of relative but positive noninterventionism can be found in the development of creative industries, which should ideally be left to private-sector groups and individuals. The role of the government is minimal, thus unleashing the potential of both groups and individuals in the growth of creative sectors. Freedom of expression, thought, and speech is of paramount importance in the growth of creative industries, whose deepening may also necessitate more intervention and support from the post-1997 government.

The role of the Macao government is slightly different. Macao, under Portuguese rule until December 20, 1999, attached great importance to cultural development, especially to the protection of Portuguese culture and heritage in the territory. Because of the Portuguese emphasis on culture and arts, the Macao Special Administrative Region (MSAR) has witnessed the proliferation of creative industries, including culture and arts organizations, with strong support from the government. Apart from the provision of government subsidies in support of creative sectors, a whole range of groups and individuals has sprung up in the MSAR—a testimony to the unique partnership between government on the one hand and individuals and groups on the other.

The ROC on Taiwan adopts a similar model. Taiwan's culture, including that of indigenous peoples residing on the island republic, can be regarded as unique. While groups and individuals have designed various strategies in promoting their creative industries, the government of Taiwan has also realized the sociopolitical importance of engendering the unique Taiwanese identity through the development of creative sectors.

Hence Greater China—mainland China, Taiwan, Hong Kong, and Macao—is witnessing three different paths of developing the creative industries. The PRC is adopting a state-centered model of development, adhering to the principles of economic modernization and political correctness in the fostering of various creative industries, which are expected to serve the interests of the state and the ruling party. The Taiwan model is similar to Macao's, forging a close partnership between government and groups and individuals while unleashing the market to produce synergy with all actors concerned. The Hong Kong model is largely market-driven, punctuated by the minimalist role of the postcolonial state—a legacy from the British colonial era, as the colonial government tended to put the development of creative industries on the back burner.

THE ROLE OF GOVERNMENT AND ITS DEFINITION OF CREATIVE AND CULTURAL INDUSTRIES IN GREATER CHINA

Comparatively speaking, the governments of Greater China— mainland China, Taiwan, Hong Kong and Macao—have varying definitions of creative and cultural industries, leading to different calculations with regard to the role of creative industries in the economy. Table 0.1 shows the different methods of calculating the role of creative and cultural industries in the economic development of the four parts of Greater China. In Taiwan, cultural and creative industries accounted for about 4 to 5 percent of overall GDP from 2002 to 2010, while the number of people employed in both cultural and creative sectors increased slightly, from 162,436 in 2002 to 195,156 in 2008. Yet in terms of the annual growth rate of Taiwanese people employed in these two sectors, the figure declined from 3 percent in 2003 to -7 percent in 2008. Hence the development of creative and cultural sectors has been mixed in Taiwan, whose government tends to mix the two in its economic consideration and calculations.

Hong Kong is similar to Taiwan in the sense that both governments amalgamate creative and cultural industries in the evaluation of GDP growth. Overall, Hong Kong has seen a gradual increase in the economic growth of creative and cultural industries, rising from 3.8 percent of GDP in 2005 to 4.6 percent of GDP in 2010. This rise is consistent with the slight increase in the percentage of people working in cultural and creative industries in the overall population (from 5.1 percent in 2005 to 5.4 percent in 2010). The number of people employed in cultural and creative industries increased from 3,343,000 in 2005 to 3,503,000 in 2010. The absolute number of citizens employed in the two sectors was larger than in Taiwan.

Why did Hong Kong have more people participating in creative and cultural industries than Taiwan, even though the population of Hong Kong is much smaller than that of Taiwan? It appears that the Taiwanese government's definition of creative and cultural industries is much narrower than Hong Kong's definition. According to the Taiwanese government, the cultural and creative industries embrace areas such as mass media, performing arts, visual arts, broadcasting media, and customs festivals.[2] The definition of cultural and creative industries offered by the Hong Kong government is far clearer and much broader: "the coverage and classification of cultural and creative industries are based on international statistical guidelines promulgated by the United Nations Conference on Trade and Development (UNCTAD) and the United Nations Educational, Scientific and Cultural Organization (UNESCO), with appropriate adaptation to cater for the economic situation in Hong Kong."[3] Specifically, according to Hong Kong's definition, creative and cultural industries "comprise 11 component domains as follows: arts, antiques and crafts; cultural education and library, archive and museum services; performing arts; film, video and music; television and radio; publishing; software, computer games and interactive media; design; architecture; advertising; and amusement services."[4]

In mainland China, the percentage of the population involved in creative and cultural industries is not revealed in official statistics; instead, the PRC focuses on the cultural industry as a percentage of overall GDP development, which was 2.15 percent in 2004, rising slightly to 2.85 percent in 2011. However, the mainland Chinese definition of cultural industries is as broad as its Hong Kong counterpart, encompassing the mass media, broadcasting, cultural and arts services (including libraries, museums, archives, and cultural heritage protection organizations), information services (Internet and telecommunications), cultural creativity and design sectors (advertisement, comics, architectural design, indoor design), leisure and entertainment services (parks management, zoos management, games centers, and photography services), and traditional arts and painting (artisan works, jewelry, carpets, porcelains, antiques).[5]

Table 0.1 Relations between creative and cultural industries and the economies of Greater China

Taiwan (2002–2010)									
	2002	2003	2004	2005	2006	2007	2008	2009	2010
Turnover by cultural and creative industries in billions NTD	4,353	5,032	5,565	5,811	5,998	6,174	6,091	5,698	6,616
Annual growth rate		15.60%	10.61%	4.42%	3.22%	2.93%	-1.34%	-6.45%	16.11%
% of overall GDP	4.20	4.70	4.90	4.90	4.90	4.80	4.80	4.60	4.90

	2002	2003	2004	2005	2006	2007	2008
Number of people employed in cultural and creative industries	162,436	167,443	185,758	195,684	207,785	211,550	195,156
Annual growth rate		3.08%	9.68%	5.34%	6.18%	1.81%	-7.40%

Hong Kong (2005–2010)						
	2005	2006	2007	2008	2009	2010
Value added by cultural and creative industries in millions HKD	52,258	57,309	65,117	63,275	63,266	77,683
Annual growth rate		9.7%	13.6%	-2.8%	0.0%	22.8%
% of overall GDP	3.80	3.90	4.10	4.00	4.10	4.60

	2005	2006	2007	2008	2009	2010
Number of people employed in cultural and creative industries	3,343,000	3,412,000	3,485,400	3,521,400	3,486,900	3,503,000
Annual growth rate		2.1%	2.2%	1.0%	-1.0%	0.5%
Cultural industry as % of overall population	5.10	5.20	5.20	5.40	5.40	5.40

Mainland China (2004–2011)								
	2004	2005	2006	2007	2008	2009	2010	2011
Value added by cultural industry (in billions RMB)	3,440	4,253	5,123	6,455	7,630	8,594	11,052	13,479
Annual growth rate		23.6%	20.5%	26.0%	18.2%	12.6%	28.6%	22.0%
Cultural industry as % of overall GDP	2.15	2.3	2.37	2.43	2.43	2.52	2.75	2.85

Macao (2008–2013)						
	2008	2009	2010	2011	2012	2013
Consumer Price Index indicator						
On recreation and culture	97.41	100.55	104.56	109.95	112.42	
Number of participants in cultural activities					263,800	285,300
Employed population						366,700
Recreational, cultural, gaming, and other service employment as % of overall population						25.40

Sources: Annual report on cultural industries development in Taiwan (2003–2011), http://cci.culture.tw/cci/cci/epaper.php?act=search_ye&ddlSearchYE-Year=2011; *Hong Kong Monthly Digest of Statistics,* April 2012, http://www.censtatd.gov.hk/hkstat/sub/sp80_tc.jsp?productCode=FA100120; mainland China statistics, http://cpc.people.com.cn/18/BIG5/n/2012/1114/c350837-19570868.

html; Macao Statistics and Census Service, http://www.dsec.gov.mo/default.aspx.

Note: NTD stands for new Taiwan dollars; RMB stands for renminbi; HKD stands for Hong Kong dollars.

Comparatively speaking, the Macao government adopts the broadest definition of creative and cultural industries, as it incorporates the gaming sector, and casino capitalism is the pillar of economic growth in the MSAR. Table 0.1 shows that participants in cultural activities amounted to 285,300 in 2013 and that about 366,700 citizens were employed. Most importantly, citizens employed in recreational, cultural, gaming, and other services made up 25.4 percent of the overall population in 2013.

Table 0.2 China's cultural and related industries classifications in 2012

1. News and publication services	News services Publication services Distribution services
2. Broadcasting, television, and filmmaking services	Broadcast and television services Filmmaking and audiovisual recording services
3. Cultural and arts services	Cultural creativity and performing services Library and archival services Cultural heritage and preservation services Mass cultural services Cultural studies and community services Cultural and arts training services Other cultural and arts services
4. Cultural news communication services	Internet communication services Value-adding telecommunication services (cultural sector) Broadcast and television communication services
5. Cultural creativity and design services	Advertising services Cultural software services Architectural design services Professional design services
6. Cultural recreation and entertaining services	Scenic touring services Entertainment and recreational services Photographing and printing services
7. Craft artwork production	Craft artwork manufacturing Landscape, decorative art, and other ceramic products manufacturing Craft artwork merchandising

Source: National Bureau of Statistics of China, http://www.stats.gov.cn/tjbz/t20120731_402823100.htm.

As a matter of fact, the Macao government has set up departments and statutory bodies responsible for the promotion of cultural and creative industries, especially since Macao is regarded as a unique cultural place for not only the local Chinese but also the Portuguese and Macanese (locally born Macao people of Portuguese–Chinese or Portuguese–Malayan ancestry). The Department for the Promotion of Cultural and Creative Industries is responsible for "helping to formulate policies and strategies pertaining to the promotion of cultural and creative industries, for facilitating and nurturing the growth of local talent to further the development of cultural and creative industries, for proactively promoting local cultural and creative product brands and building a business environment conducive to the growth of cultural and creative industries, for conducting studies on the developmental orientation of Macao ... and for informing the public of the status of these industries."[6] Moreover, the Macao government has approved establishment of the Macao Creativity Association, an independent legal and nonprofit entity providing a platform for the training of creative talents, the generation of ideas for stimulating the development of the creative industry, and the submission of policy proposals to the government.[7] The association includes individual and corporate members spanning different occupational sectors, including tourism, conventions and exhibitions, fashion, design, cultural media, information software, gaming and entertainment, advertisement, publications, construction and land development, the arts, antiques, and scientific research.

Finally, the private sector has also set up research institutes for the promotion of creative industries. For example, in 2003 the Center for Creative Industries was founded by the Institute for European Studies of Macao with the objective of training talent for the creative industries through workshops, educational seminars, conferences, and the sharing of ideas and networks. The creative areas identified by the institute include advertising, architecture, crafts, design, fashion, film and video, interactive leisure software, music, the performing arts, publishing, software and computer development, and visual arts.[8] Overall, the Macao government attaches great importance to the broad definition and scope of creative and cultural industries to promote the territory's economic prosperity and to forge close relations with the casino and tourist sectors.

In short, all four places in Greater China have varying definitions of creativity. All four governments tend to link creativity with cultural industries, but the Hong Kong definition of creative and cultural sectors tends to be as broad as that of the PRC, while the Taiwanese definition is much narrower. The Macao definition turns out to be the broadest, mainly because the government combines creative and cultural industries with the casino and tourist sectors to maintain and expand the momentum of economic development in the relatively small territory.

Chapter 1, written by Michael Keane, focuses on the debate within the PRC about an apparent growth surge in China's cultural and creative industries. The PRC government is serious about accelerating its global branding strategy under the rubric of "soft power" and is throwing significant resources into media and culture sectors, arguing that culture will inevitably become a "pillar industry" (*zhizhu chanye*). Keane assesses how China's creative industries might learn by building better relationships with their East Asian neighbors. He describes a six-stage timeline through which China's creative industries hope to become competitive regionally and even internationally. Keane draws attention to the main problem facing mainland China: too much focus on industry and not enough on the creative dimensions. It is in this critical area where East Asian talents, managerial capability, and technological expertise can play a crucial role in making the mainland's soft power more attractive. Keane's important chapter highlights the role of government in the development of creative industries, especially the utilization of Chinese culture to facilitate economic growth rather than to achieve creativity. Here, the socialist ideology in the PRC and its combination of economic growth and creative industries are noteworthy.

Chapter 2, written by Xuan-Olivia Jiang, stresses the importance of creative industry policy transfer from one city to another—in particular, what the cities of Guangzhou, Beijing, and Shanghai can learn from one another in the process of promoting the development of creative industries. Her chapter has significant implication for our analytical framework of understanding creative industries because the level of government and its learning do matter. The municipal level of government in China and its policy learning and transfers contribute to the development of creative industries.

Anthony K. C. Ip's chapter 3 focuses on the cities of Chengdu and Kunming in embarking on cultural revitalization and pedestrianization in their urban cores since 2000. In these historic cities, the conventional roles of enclave, haven, and outpost have been transformed into that of regional economic-technological hub and Southeast Asian gateway. The development of Chengdu and Kunming demonstrates innovations in high-level industrialization, education, mineral resources and precious-metal extraction, biotechnology, and urban lifestyle. The synergistic nature of transforming new development paradigms and investments is reflected by capitalizing on the rich Han and minority cultures. This chapter profiles an area called Wide and Narrow Lanes and other old streets, analyzing their design and planning within social, economic, technological, and policy contexts as foundations for further research and development. The chapter highlights the role of city design and planning in the development of creative industries in China.

Jane Zheng's chapter 4 delineates artistic creativity as a key issue in art education. This chapter explores ways in which creativity was fostered in art education at Shanghai Art College during the Republican period. It draws upon several primary sources, such as school archives and artists' memoirs, to show two levels of creativity in the school's art education. At the school's administrative level, creativity was advocated in line with the policy of the PRC Ministry of Education and referred to loose discipline in class and adequate freedom for the learning and expression of students. At the level of instruction, the literati's way of artistic creativity was retained and applied to a wide range of painting genres across cultures, thus producing a new artistic form that mixed Western with Chinese painting components at the spiritual level instead of integrating techniques and styles into representation. Her chapter clearly focuses on the role of schools and individuals in fostering creativity in China.

Minxing Zhao's chapter 5 focuses on the role of the PRC in controlling and combating cultural heritage crimes. Again, the role of the government is critical to the preservation and development of creative industries, which must utilize cultural heritage as a means of attracting tourists. However, many cultural heritage sites have been damaged by criminals, and to protect cultural heritage, a legal response is necessary. Still, the theft of cultural relics in the PRC remains rampant.

Chapter 6, written by Ming Lai and Chi Cheung Leung, analyzes the cultural discourse of the government of the HKSAR through its policy addresses from 1997 to 2013. Using keyword analysis, the authors find that the term *culture* was sometimes replaced by *creative* and *industries* and that these three terms appeared simultaneously at times. Moreover, the term *arts* disappeared for two consecutive years, from 1999 to 2000, but has appeared again in the discourse about cultural and creative industries and the West Kowloon Cultural District. In his policy address, Chief Executive C. H. Tung (1997–2004) expressed that the hallmark of Hong Kong culture was to act as a bridge between East and West. Moreover, the Hong Kong people, according to Tung, should be more knowledgeable about the culture of the motherland. The cultural discourse of the second chief executive, Donald Tsang (2004–2012), is summarized as having 'two pillars plus many concrete projects." The two pillars were the West Kowloon Cultural District and cultural and creative industries. In his first policy address, the third chief executive, C. Y. Leung, reiterated the topics of East–West cultural exchange and various cultural projects. Clearly, the Hong Kong government is instrumental in developing creative industries through infrastructure projects and policy pronouncements.

Chapter 7, by Wai Shing Lee, provides a critique of Hong Kong International Museum Day (HKIMD), an annual event organized and subsidized by the HKSAR government. He questions the effectiveness of a promotional campaign that has lasted for more than ten years under the management of Leisure and Cultural Services Development. Lee argues that visitor attendance rates may not reflect the usage of museum facilities. In fact, HKIMD attained high attendance figures when holding blockbuster exhibitions. However, the content of educational and public programs was found to be too entertaining in nature and not linked to studies of art forms in the art curriculum. The campaign cannot increase the public's appreciation of cultural heritage if programs are too leisure oriented and entertaining. The original mission of running HKIMD was not really achieved, suggesting an urgent need for review of its policies and content to engage more public support for museums and their activities. In brief, museums as public organizations have to critically assess their own activities and reexamine their roles in developing cultural heritage and its industries.

Based on a rich historical context in Hong Kong and mainland Chinese history, in chapter 8 Roger Ting Sing Kho observes that numerous key Chinese calligraphy writings have been dispersed in public spaces in Hong Kong—a reflection of recent transformations of architectural design. The calligraphy works have been carved on architectural facades in monuments, and they define the character of the city, forming a key mental map and opening a pathway to the display of living heritage as a symbol of common Hong Kong identity. The art form has reflected Chinese culture for centuries under a melting pot of different cultures, politics, and religious beliefs. As a methodology of inspecting and polishing image through Chinese characters, calligraphy is a powerful tool, enabling viewers to decipher artistic style and promoting art appreciation. The methodology is accessible to the public, and its architectural uniqueness can preserve key remains of a city's cultural heritage under rapid urban transformation. Kho emphasizes identity creation through public artworks produced by individual artists. Hence the role of individuals is stressed in the development of Hong Kong's cultural industry.

Yan Tung in chapter 9 analyses successful examples of cultural enterprises, including public and private ones, from an art practitioner perspective. She focuses on venturing strategies during their founding years, with an ultimate objective of identifying feasible tactics. These tactical measures embrace first the application of geographical proximity and place-making strategies, meaning that enterprises clustering in a region without art can save on rental costs and differentiate themselves from the rest, especially if they can build up a unique art identity. Second is an intelligent and skillful use of promotional strategies, including online tools and social media channels through which staff can broaden membership bases. Furthermore, enterprises can foster networks within and across the arts field, generating leadership. The third tactic is partnership with parallel cultural organizations, such as cultural institutions, for the sake of relatively stable income and participants. Finally, mobilizing and galvanizing financial capital in multiple cities outside of Hong Kong has proven to be effective in promoting international branding and trade. In a nutshell, Yan Tung emphasizes the strategic dimensions of cultural enterprises in the promotion of creative industries.

Based on his unique theoretical framework on the development of cultural entrepreneurship, Chi Cheung Leung in Chapter 10 assesses the features of four community orchestras in Hong Kong. His study investigates the dynamic relationship between artistic leaders, management, and cultural entrepreneurs. He argues that the space for cultural creativity is actually tremendous, depending on individual leadership, organizational or group management, and strategic planning as a whole. Leung's chapter provides a perspective for us to comprehend the various actors, institutions, groups, and individuals in the complex process of developing creative industries. His study highlights various strategies in developing entrepreneurship. He identifies four models: the new-generation model, the affiliation-based model, the mentor–mentee model, and the developmental model.

As described in chapter 11, by Chiu Yu Mok and Eric Ng, the Centre for Community Cultural Development (CCCD) in Hong Kong was founded in 2004 by Mok and like-minded individual artists. Its objectives are to promote and practice community cultural development, including cultural activities to empower the underprivileged, in particular persons with disabilities, migrant workers, and ethnic minorities. The chapter focuses on an independent artistic group in Hong Kong. The writers are the forerunners and founders of CCCD. They present a detailed account of the organization and its founding motto, which is firmly entrenched in the community-building process and is punctuated with social change campaigns to advocate for diversity in the cultural development of Hong Kong and the social renaissance of the PRC. Early modern China used performative art forms such as theater and song to celebrate the peasant working culture and ideologies. The CCCD was founded on the ideology of empowering people to assert and express their rights through art education as depicted by UNESCO standards. It organizes workshops and performances in cross-disciplinary areas such as film, visual art, dance, community music, and art therapy. In short, Mok and Ng operated the CCCD as a crucial cultural platform in the creative expression of social, political, and cultural visions for both Hong Kong and China. Their emphasis on creativity is constructed on the basis of individual determination, perseverance, and group dedication.

Chapter 12, by Vivienne Chow, centers on the licensing policy of the Hong Kong government, including procedures of the governing broadcast authority involving free TV licenses. She argues that the availability of free TV entertainment channels is an indispensable factor in maintaining the cultural identity of Hong Kong's people. To Chow, the cultural values and diversity of Hong Kong are represented by the existence of a free and open media environment, especially if the mass media serves as an intermediary between policy makers, content producers, and the public. Chow argues that Beijing appears to be a crucial actor behind the scenes of the free TV licensing saga and that the core values of rule of law and freedom of expression might be threatened. Hence she focuses on the role of both government and groups in maintaining cultural freedom and creativity.

In chapter 13 Victor Kwong investigates creativity in Hong Kong in the past decade by examining talent, technology, and tolerance: three factors discussed by Richard Florida. Kwong finds that, comparatively speaking, Hong Kong has performed quite well in Asian rankings and that ways of retaining the creative class in Hong Kong and expanding the creative potential of Hong Kongers can be studied further.

In chapter 14, Joey Moon compares and contrasts Taiwanese cinema with it Hong Kong counterpart. She examines the extent to which the fluid character of the *flâneur* is explored in Taiwanese and Hong Kong cinema. The concept of the *flâneur* is used as an empirical framework to interpret reality, as well as to examine the particularities of individuals who symbolize fluid boundaries between locals and outsiders according to the principles of cosmopolitanism. Two films that portray these contemporary social realities are *Parking* (2008) by Mong-Hong Chung and *Sparrow* by Johnnie To (2008). The *flâneur* characters portrayed in these films can be divided into two social groups—namely, locals and outsiders who moved from China to either Taiwan or Hong Kong. They symbolize new fluid boundaries that are redefining the open societies of contemporary Taiwan and Hong Kong. In addition, they articulate connections and contradictions among Taiwan, Hong Kong, and China. The new type of *flâneur* highlights the contrast between locals and outsiders and negotiates cities that are more fluid. These contemporary cities allow characters to experience a range of psychological possibilities, such as rejection, questioning, confrontation, and acceptance in urban landscapes. In short, Moon focuses on individual characters in cinema as an expression and articulation of creativity.

Chapter 15, by Patrick Jones, focuses on the education of entrepreneurial portfolio musicians in the twenty-first century. Adopting a global and comparative perspective, Jones utilizes the example of Syracuse University to illustrate a curriculum that emphasizes the development of the teaching, performing, composing, and other entrepreneurial skills required to sustain a musical career. According to Jones, university students who specialize in music need to develop their teaching, performing, and composing work. Syracuse University designs courses and internship projects so that both American and Chinese students have tremendous opportunities for local and international involvement in research, exchange, collaboration, production, and projects. Students are exposed to differences between Eastern and Western schools of cultural thought and the need to incorporate global perspectives, especially if Chinese musicians would like to reach out to more international audiences.

Chapter 16, by Rita Yip and Ji Hong Ye, examines the inclusive cultural policy in the music education of Nanjing and Hong Kong. They adopt indigenous insider and outsider perspectives in analyzing the enacted curriculum in primary and secondary school music education in both cities. They examine a range of materials, styles, and musical forms, including ethnic folk songs, children's songs, nursery rhymes, instrumental music, Western classics, and popular songs, combined with the use of images, sound clips, and audiovisual materials. Yip and Ye found that singing and listening materials are more often employed than playing materials, in particular with instrumental music, suggesting the need for professional development in the training and knowledge of music teachers. Western classics are preferred to other world and pop music, suggesting the need to include more contemporary music and other innovative sounds in a more culturally balanced curriculum to maintain an open approach in addressing diversity in music education. Yip and Ye conclude that Hong Kong employs more Western music than Nanjing and that both cities should ideally utilize more global and cross-cultural music, especially in music education.

Chapter 17, by Patrick Yuen, examines the impact of market-driven phenomena on the film industry in Hong Kong, which as witnessed the growth and struggle of coproduced films in a culturally heterogeneous environment. Using figures from multibillion-dollar mainland box office revenues for the last seven years, the chapter explores opportunities and restraints within the changing political and ideological context, including how Chinese censorship and foreign investment policy affect the filmmaking industry, on top of commercial and mainland-audience influences on directors' cinematic style. The collapse of the Hong Kong TV and movie industry has facilitated the integration of Hong Kong into the mainland Chinese film industry, which is undergoing the transformation of the PRC's nationalistic film industry to one dominated by market-determined productions and subsequent regulations on film content. The chapter describes further reactions from the globalized market, such as production of a series of test films to obtain approval of Chinese censoring authorities on sensitive areas of ideological and political correctness. Discussions include how the PRC government controls the influx of Hollywood films by imposing a strict quota system and requirements to employ Chinese workers on foreign projects, resulting in the adaptation of scripts and the integration of Chinese cast members, producers, and directors into international productions. In short, Yuen focuses on the dynamic interactions between the Chinese government and Hong Kong producers and filmmakers.

Chapter 18, by Sonny Shiu-Hing Lo, focuses on the role of political comics artists in protecting and maintaining freedom of expression in Hong Kong. In contrast, mainland Chinese comics artists tend to adhere to the principle of political correctness. Lo argues that while freedom of expression shapes the prosperous development of political comics in Hong Kong, comics critical of the PRC regime are rare, and comics adhering to the principles of political correctness, such as anticorruption, are emerging in mainland China.

All the chapters in this book provide in-depth analysis of the role of governments, individuals, and groups in the development, growth, and transformation of creative and cultural industries. Some contributors focus on the role of individuals; some on groups; some on government policies. Some combine one or two of the three crucial actors. Moreover, the role of the market is essential in sustaining the development of creative and cultural industries, with or without the financial support and interventionist policy of the government. All the chapters contribute to a deeper understanding of the dynamics and interactions between government, the market, groups, and individuals in the birth, evolution, and consolidation of creative and cultural industries.

NOTES

1. See C. B. Macpherson, *The Real World of Democracy* (Oxford: Clarendon Press, 1966).
2. For this broad definition of creativity, see Michael Michalko, *Creative Thinking* (Novato, CA: New World Library, 2011). Also see Jason Potts, *Creative Industries and Economic Evolution* (Cheltenham, UK: Edward Edgar, 2011), and Terry Flew, *Global Creative Industries* (Cambridge: Polity Press, 2013). In *Management and Creativity* (London: Blackwell Publishing, 2007), Chris Bilton writes, "Creativity is everywhere and nowhere—paradoxically, while it is accessible to all, it is nevertheless marketed as a rare commodity."
3. See *The 2009 Cultural Statistics of Republic of China* (Taipei: Government of the Republic of China on Taiwan, 2009).
4. See *Hong Kong Monthly Digest of Statistics May 2013* (Hong Kong: Census and Statistics Department, Hong Kong Special Administration Region Government, 2013), FB3, section 2.2.
5. Ibid., section 2.3.
6. See the categories of "cultural and related industries" at the website of the National Bureau of Statistics of the People's Republic of China, http://www.stats.gov.cn/tjbz/t20120731 402823100.htm (accessed December 6, 2012).
7. See the website of the Department for the Promotion of Cultural and Creative Industries, http://www.icm.gov.mo/en/Structure/StructureMain.aspx?id=12 (accessed December 12, 2013).
8. See http://www.mcia.org.mo/mcia/introduction?tabIndex=2&rightTabIndex=1 (accessed December 12, 2013). See http://www.creativemacau.org.mo/EN/overview.phd (accessed December 12, 2013).

Section 1: China

Chapter 1

Scrambling for the Mainland China Market

Michael Keane

China's growing economic, political, and military capacity is arguably the most geopolitically significant development of the twenty-first century. With the oldest recorded written civilization in the world, China is home to many cultural treasures. But internationally, is China's culture admired or just respected? Is "brand China" attractive? In 2011 former president Hu Jintao lamented China's cultural weakness. He said, "The overall strength of China's culture and its international influence is not commensurate with China's international status. The international culture of the West is strong while we are weak."[1]

How then should China proceed? Should it learn from the West? Should it adopt strategies that have made the developed economies so successful, generating talent and technology clusters like Hollywood and Silicon Valley? Indeed, it is not uncommon to hear of attempts to "upgrade" China's economy by transplanting international development formulas. The city of Hengdian in Zhejiang Province has for several years advertised itself as China's Hollywood, or Chinawood, despite a business model that relies on low-budget production of TV serials and movies.[2] In 2013 the second-tier city of Zhangjiang in coastal Jiangsu Province announced plans to "become China's Silicon Valley." The party chief of Zhangjiang, Yao Linrong, visited the United States in April 2013 and noted, "Talent equals technology and we need to bring in more high-tech talent to create the climate for good innovation."[3] What is striking is that such international technology and talent transfers are facilitated by Chinese Communist Party leaders and officials. Equally significant, however, is the fact that the term *pillar industry* refers to sectors that receive state protection and subsidy.⁴

In 1956 Chairman Mao Zedong, often associated with China's disdain of markets prior to the 1980s, remarked, "In the industrially developed countries they run their enterprises with fewer people and with greater efficiency and they know how to do business. All this should be learned well in accordance with our own principles in order to improve our work."⁵ The key sentence in this well-known citation from *Contradictions* is "They [the foreigners] know how to do business." Certainly, Hong Kong entrepreneurs know "how to run their enterprises." Following the Chinese civil war of the 1940s and the Japanese invasion of the mainland, hundreds of thousands of refugees and migrants resettled in the British colony, eventually positioning Hong Kong as a financial hub and, by the latter decades of the twentieth century, making it a conduit for trade and investment in the mainland. Hong Kong film and TV industries achieved remarkable regional and international success in the 1980s and 1990s, although recent times have seen a move to the mainland in search of bigger audiences. From the heyday of its film industries—Golden Harvest's "discovery" of Bruce Lee and Jackie Chan—to Run Run Shaw's Movie Town and Shaw's eventual move into television, Hong Kong has been an incubator of talent and a distribution center of Chinese culture.⁶

In this chapter I show how China's cultural and creative industries are evolving and why a deeper and more holistic understanding is important. In some cases China's creative sectors are becoming competitive; however, in the main they are struggling to break out of the straitjacket of pedagogy. Based on interviews and fieldwork in China over the past decade,⁷ I show what this ecology looks like to a participant—for instance a start-up company, entrepreneur, artist, or university graduate seeking to move into cultural production. Following this I make some observations about how East Asia's creative industries fit into the picture vis-à-vis the mainland.

In short, many power brokers in China want the nation to become a powerful cultural exporter—that is, a "strong cultural nation."⁸ The problem is that China doesn't quite know how to transform its production model. To put it more succinctly, it doesn't have the requisite know-how; for instance, the kind of creative know-how that made Golden Harvest such a success. Success also characterizes the rise of South Korea where the *chaebol*, a corporate form of the state-owned enterprise model, has provided stability and financial muscle. In Korea cultural and content industries have found their creative edge; in the past decade has arisen a sense of imagination and risk-taking that was conspicuously absent under previous authoritarian regimes. Taiwan's robust democratic system has also fanned the flames of creativity and provided market leadership in pop music. In short, the mainland needs East Asia for its knowledge of markets as much as East Asia now looks to the mainland as a marketplace and a source of new talent. Indeed, markets for Chinese creative content are growing as China's population urbanizes and as the state looks for ways to realize what it calls "cultural soft power."[9]

YOU TAKE THE HIGH ROAD, BUT WE'LL TAKE THE LOW ROAD

I explain China's ascendency as a strong cultural nation through the heuristic of a "cultural innovation timeline." In this timeline I highlight the nature of industrial upgrading and value adding. My basic premise is that a firm or a business "chooses a point in economic space as its test of market conjecture, and then commits resources to the choice."[10] Although the timeline model applies most directly to start-up businesses, it also describes movements of human capital; for instance, designers moving into emerging animation sectors; animation graduates moving into games industries; or, in the case of Dafen Painting Village in South China, migrant laborers taking up oil painting.[11]

The cultural innovation timeline is a qualitatively different approach than national innovation systems (NIS) or regional innovation systems (RIS).[12] These terms apply to innovative capacity (IC) and draw heavily on indicators such as the number of high-tech firms, the number of patents filed, and the number of students graduating with PhDs. The timeline avoids these abstract determinants of innovation and looks at how artists, designers, and media practitioners are entering the market at particular points in time.

The timeline model illustrates the aspirations of the mainland to be a player in "soft power competition" and a "strong cultural nation."[13] The timeline illustrates how China's arts, media, design, and cultural businesses are looking to move up the value chain. In spite of this impetus, most government policies are still directed at low-value segments: it seems that producing low-cost components for international markets influences many of the decisions pertaining to investment. The economic geographer Wang Jici has aptly termed this approach "taking the low road."[14]

How might participants "take the high road?" Indeed, how many are choosing to commit resources, to experiment, to take risks? Factories, bases, studios, and cultural parks are busy producing artifacts and commodities, as well as business, consumer, and tourism services. As participants move up the value chain, we note more evidence of collaboration and partnerships, both cross-platform in a technological sense, across different industrial sectors, and across cultures. The timeline illustrates coevolution in arts, design, and media activities. In the upper levels of the model, there is a greater focus on intangible services and virtual business models, ensuring considerable creative destruction.[15] Overall a distinctive form of postindustrial transformation is evident, moving China's arts, design, and media industries into the intangible or "weightless economy" typified by the term *creative communities*.

INDUSTRIAL UPGRADING

With a steady boom in China's economy over the past two decades, mostly due to low-cost production and export processing, it is clear that market activity is well organized. How does this apply to creative fields? What characterizes creative production—and consumption—in China today?

China's efforts to become an innovative nation have received a great deal of attention in the media. In *How Creativity Is Changing China*, Li Wuwei writes, "From the perspective of industry development there are two ways to stimulate the transformation of China's economic growth. One is to increase the value add of China's manufacturing industries through upgrading, readjusting and innovating in terms of industrial structures. The other way is to develop new, innovative, eco-friendly industries with growth potential"[16]

The word upgrading (shengji升级) occurs frequently in Chinese industrial economics and in policy documents. The role of policy is to provide the right levers to drive accumulation and upgrading of the economy, to encourage investments in clusters and infrastructure. In economic theory this is essentially a neoclassical model with socialist underpinnings. However, transforming from manufacturing to creative services entails more than just upgrading and adjusting. The third element in Li's prescription, innovation, is more elusive.

The challenge of innovation implies a connection with a more Schumpeterian approach in which entrepreneurial activities generate structural change and produce uncertainty. Creative endeavors are based on uncertainty—that is, we generally don't know whether we like a cultural product until we consume it. While future return on investment in creative endeavor is often difficult to predict, this does not mean that the environment for creative activity cannot be planned or even stimulated.

The issue of what counts is a contentious topic. Evidently, while many artists, practitioners, and businesses in China aspire to develop original works, there are mechanisms that constrain originality and policies that reward replication of state-approved forms. This is a legacy of the authoritarian state in which the role of culture was to enlighten rather than to stimulate curiosity. In the current environment the focus has turned to productivity: how to make culture a "pillar industry." The challenge can be expressed as follows:

China is producing more cultural goods and services as its population seeks more diverse consumption opportunities. Cultural goods and services are increasingly produced with business and consumer markets in mind.[17] Despite state-imposed restrictions, China has imported increasing levels of cultural products from Taiwan, Hong Kong, and Korea over the past decade. These imports have assisted Chinese producers to develop better-quality products, although, as Chua Beng-Huat maintains, China is essentially a consumer of East Asian culture rather than an exporter.[18] The ignominy of being an export destination has generated concern about China's cultural sovereignty. The challenge of ensuring "national cultural security" (guojia wenhua anquan 国家文化安全), first raised in 2000, was followed in 2005 by reports in the People's Daily of a "cultural trade deficit" crisis (wenhua maoyi chizhi 文化贸易持之).[19]

By 2007 China's leaders had determined that its culture should go abroad, to build and extend the nation's soft power. State and local governments offered incentives to people to join the new workforce, to move into clusters. It was now a patriotic duty to be creative. For people who had previously worked in state-financed cultural troupes or publishing houses, the challenge was now to form private enterprises (qiye 企业). But what does this window of opportunity mean for start-up companies, for entrepreneurs, for small-time media professionals? Where do they enter into the market? At what point in economic space do they commit resources? How do political and social networks impact on such choices?

I identify six stages or levels in the development timeline of China's creative economy. To some extent these overlap and interact, but they can also be regarded as discrete development trajectories. China is transforming from its strengths in material culture (artifacts, goods, performances) toward the intangibles of the digital economy, one in which markets are more complex and fragmented. In short, my underlying premise is that creativity needs time. Regardless, Chinese policy makers want to fast-track progress, to accelerate cultural development as the nation has reportedly done in technological innovation.

The six levels illustrate catch-up strategies and explain to some extent why Chinese policy makers are prescribing industrial models for culture. They reveal the insecurity of China's soft power—namely, the ignominy of being typecast as a derivative nation by East Asian neighbors South Korea and Japan, as well as by the Special Autonomous Region of Hong Kong and Taiwan, which have harvested the gains of commercial contemporary media.

A fundamental challenge facing China, therefore, is how its cultural workers, artists, and designers can generate "original" works and performances rather than providing technical services, particularly to entities that are non-Chinese? This is also the challenge of moving from simple cultural goods to complex cultural goods, the latter category entailing greater complexity of financing and sharing of revenues.[20] In addition, moving beyond the "world factory model" entails a deep commitment to allowing more imaginative ideas.

Table 1.1 The cultural innovation timeline

Stage/Theme	Strategic Form
Standardized production	Subcontracting (fashion, animation, software, toys, furnishings, electronics)
Imitation	Import substitution, local versions, and cloning
Collaboration	Coproduction and various forms of sharing knowledge
Trade	Beginning of soft power strategy; breaking out of domestic constraints
Clusters	Attempts to harness soft power by industrializing culture
Creative communities	Borderless social network markets; reaches domestic and international online audiences

Note: See Keane, *Created in China;* Keane, *Creative Industries in China.*

My main concern is to draw attention to level three (collaboration) and level six (creative communities). Collaborative practices represent the kinds of activity that are most amenable to players from East Asia.[21] Creative communities in turn are generating new forms of content that draw on East Asian influences.

STANDARDIZED PRODUCTION: THE INDUSTRIAL WORLD

The first of the six levels is standardized production—the ubiquitous "Made in China." This is the most direct illustration of how the factory model is *transferred* into the domain of culture, the world of outsourcing—fashion sweat shops, animation rendering, and toy making. For the past decade this "industrial world" of standardized production has driven China's economic success in technological components, household appliances, medical equipment, tools and machinery, and plastics—virtually every commodity found in supermarkets and hardware outlets globally. Major international brands have established factories to take advantage of low rent and cheap labor. Production-line labor in these factories is managed by strict routines.

The standardized production model is well illustrated in fashion, industrial, and product design. The creative element is supplied by the international firm while the production is done in China. The brand label may claim that the product is designed in the home country, but increasingly one finds a second production label, "Made in China." In many instances products are anonymous, unbranded, shipped to a number of buyers—for instance, Walmart. In other instances international companies such as IKEA set up managed production centers in China with highly regimented routines and careful protection of "trade secrets."[22]

The weakness of the standardized production model is that it perpetuates a bottom-line mentality. The availability of cheap land for factories and clusters (see level four) has implications for innovation. This applies to Hong Kong companies that have based their businesses on exploiting cheap labor in South China. The strategy of focusing on an industrial model of production with a high proportion of OEM (original equipment manufacturer) and low-cost technology runs the risk of contract work moving to lower-cost locations.[23] A problem with this model therefore is that China is locked into processing, with little or no money invested in R&D. Other countries and regions in East Asia have built-in strategies to move up the value chain by developing OBM (own-brand manufacturing) capabilities or by moving to "complex" activities that require more highly skilled participants.

IMITATION: LOW RISK-TAKING

Copying is intrinsic to all learning activities and engenders novelty, which is in effect differentiation with some intrinsic goal. Local versions account for a great deal of audiovisual content. However, when copying becomes too obvious and too widespread, audiences lose interest. A common complaint directed by audiences at Chinese film and TV drama producers concerns "cloning" (*kelong*).[24] However, imitation is a legacy of socialist cultural policy inherited from the 1940s revolutionary era, which privileged the reproduction of "model characters."

The Chinese propensity for imitation has a poor reputation internationally; international consumers are constantly reminded of the Chinese predilection for knock-offs; Chinese people are keenly aware of the dangers of copy culture. In October 2005, following a documentary shown by national broadcaster China Central Television (CCTV) on the escalation in value of reproductions of renowned ceramic masterpieces, ceramists from Jingdezhen, a production center for antique ceramics in Jiangxi Province, began to compete to produce imitations. Maris Gillette tells of how copying and counterfeiting dominated Jingdezhen's porcelain production in 2005, concluding that "A set of ideas about markets and a specific organisation of production encouraged ceramists to copy and counterfeit in search of profit. Industry workers viewed market relations as impersonal, market actors as privatised, and market activity as dishonest and potentially extremely lucrative."[25]

COLLABORATION: IN SEARCH OF NEW IDEAS AND SKILLS

The third stage is collaboration. Dissatisfaction with imitation, uniformity, and blandness leads to the quest for differentiation. This may be adaptation of a different medium or new ways of presenting a work in the same medium using new digital technologies. In China the search for market success has led many to collaborations with "foreign" partners; in many cases these partners come from Taiwan, Hong Kong, and Korea. For instance, Korean filmmakers Yi Chi Yun, Park Yeonjin, Kyonghee Noh, and Kim Jeong-Jung, all of whom studied with members of the sixth generation of Chinese filmmakers at the Beijing Film Academy, have been active in China, building personal networks across the two nations.[26]

In 2012 China's DMG Entertainment initiated collaboration arrangements with Walt Disney to produce the next installment in the *Iron Man* film franchise in China. Collaboration accelerates learning. Of course, such arrangements require central government approval. In this case, the move coincides with China's ambitions to grow the quality of its cinematic output. The form that collaboration takes varies. In film, TV, and animation industries, the preferred form is coproduction; this can be joint, assisted, or commissioned production. (This is discussed in more detail in chapter 17.) Other arrangements include Chinese versions of international magazines (*Elle, Vogue*), syndication of content, equity joint ventures, and cooperative joint ventures in advertising. The introduction of international, often Western celebrities sometimes allows Chinese brands to claim the high ground. For instance, in October 2013 the Chinese computer electronics company Lenovo Group Limited unveiled its new international designer, Hollywood actor Ashton Kutcher. According to reports, Kutcher would provide input on "design, specifications, software and usage scenarios."[27] The link with Kutcher's enigmatic portrayal of Steve Jobs in the biopic *Jobs* seemed to be a convenient way of building brand association with global market leader Apple.

Collaboration, along with foreign investment, is meant to provide benefits for both parties; in many cases, mutual advantage occurs. When one party is downgraded to a technical or production capacity, however, the result is often akin to exploitation. Certainly this has occurred in film and TV drama production when the foreign entity, often South Korean, Taiwanese, or Hong Kong, has assumed the role of creative executive producer.[28] In recent years many entrepreneurs from Hong Kong, Taiwan, Korea, and Japan have moved into media and design companies in China. Their roles are creative, managerial, consulting, and technical, providing professional expertise, alternative approaches to human capital management, and new ways of solving problems. These persons act as intermediaries, bringing ideas, investment, technology, and know-how into the sector. In addition to the East Asian cultural business migrants, many Chinese natives are returning home with overseas experience and determination to form their own companies. Policy makers are allowing these media entrepreneurs to generate ideas, to offer solutions to revitalize stagnant Chinese productions. The hope is that an increase in domestic quality, brought about by the infusion of creativity and technology, may counter the "cultural exports deficit."[29]

There are several reasons why film companies and related talent from East Asia are moving to the mainland. First, production in China can be a stepping stone to the global market. Second, economic decline in Hong Kong and Taiwan content industries is driving "creative migration," especially to places like Beijing and Shanghai. Third, media production on the mainland is relatively cost-effective. Fourth, preferential business policies, plus an availability of human capital (especially technical resources), make the mainland an attractive destination. Fifth, market entry costs are lower than in Hong Kong, Taiwan, and Korea. And sixth, the benefits of cultural proximity and shared "Asian values" can compensate for political differences.

TRADE: THE RECOGNITION OF NEW MARKETS

Trade is the fourth stage. Trade is driven by demand for new goods and services. Imports and exports of cultural goods stimulate competition. Markets for China's cultural goods, especially cinema, TV, and animation, are primarily in East Asia, despite the national government's desire to challenge U.S. entertainment and news conglomerates on the international stage. While trade is ostensibly a level beyond collaboration, it often occurs as a result of collaboration.

Trade is effectively cultural exchange; sophisticated imports trigger local consumption, which often helps local producers improve performance. China's exports to the world currently perpetuate an image of a world factory churning out cheap derivative products. Efforts by the state to increase soft power by establishing international television channels are likely to be counterproductive unless the quality of programming improves substantially. Currently, overseas CCTV channels carry signals in English, Arabic, French, Spanish, and Italian, but the schedules are filled with melodramatic, poorly made TV series and propagandistic documentaries. The global expansion of Confucius Institutes, with their emphasis on teaching language and traditional wisdom, likewise provides an impression of an unimaginative nation rather than a creative one.

CLUSTERS: COLLECTIVE REORGANIZATION OF THE INNOVATION PROCESS

The fifth stage began a decade ago, when China's leaders decided that the nation needed to industrialize culture. Following the widespread success of industrial clusters, the cluster model was grafted onto cultural production. Scores of parks, bases, and zones sprang up, often in old industrial sites. The number now runs into many hundreds, even thousands depending on elasticity of definitions. The cluster is regarded as the panacea with which to catch up, a means to fast-track development. These "spaces" of production replicate the industrialization of Chinese society over the past decade; in instances such as Beijing's 798 Art Zone and Shanghai's Tianzifang, they provide opportunities for consumption.[31]

The verb *cluster* connotes a sense of attraction—more specifically, of things, groups of people, or similar kinds coming together. Cities are in effect large clusters of diverse groups of people. Art communes are more specialized clusters of people with an interest in painting. In the contemporary Chinese usage, *cluster* connotes an advanced form of economic integration. Originating in Alfred Marshall's "industrial districts,"[32] clusters gained significance in light of Michael Porter's influential *The Competitive Advantage of Nations*.[33] Policy makers in China believe that clustering is a mechanism to fast-track economic growth, in turn realizing a range of other benefits, such as employment opportunities, real estate appreciation, and brain gain.

While the cluster momentum has taken place in cultural sectors most decisively since 2007, there are noteworthy precedents. Clustering is in effect a form of collectivism, albeit one driven by economic interests rather than social solidarity. Whereas clusters in developed countries have tended to specialize in higher-value niches, those in developing countries such as China serve segments of the market where competitiveness is determined by price.

Spurred on by the consumption-driven success of high-profile art districts like Beijing's 798 and Shanghai's Tianzifang, local officials and developers have joined forces to conjure up "creative" titles that reflect the reimagining of postindustrial space. A short list includes Creative 100 (Qingdao), Qinghai Creative Island (Dalian), Creative Warehouse (Shanghai), Creative Industries Ideas Warehouse (Tianjin), Xixi Wetlands Creative Park (Hangzhou), Shenzhen F518 Creative Fashion Park, Nanjing City of Stone Creative Park (Nanjing), Hengqing Creative Island (Macau), Foshan Creative Industries Park (Foshan), and East Chengdu Creative Music City (Chengdu).

CREATIVE COMMUNITIES: CREATION AND RE-CREATION

The last level is the most important in an evolutionary sense and in respect to how China might become a strong cultural nation. But it is the least understood. The surrounding informal environment incubates a great deal of innovation: Bulletin Board System, user-generated content, file sharing, and so on. This is the realm of informal soft power that envelopes the system; its core purpose is micro-productivity. It takes many forms—some fly-by-night, elusive, and on the border of legality. This is ostensibly household production; that is, it occurs in people's homes or workshops using new media affordances and editing software. The massive explosion of micro-films—short-form productions that are frequently experimental and genre crossing—is evidence of the propensity of amateur communities to contribute creatively to the burgeoning mediasphere in ways that are not possible for producers working in traditional media institutions such as TV networks.

The role of creative communities has come into greater focus with the growing importance of copyrighted "professional" content.[34] The

Chinese media market is encountering the pressure of continuously escalating acquisition costs due to regulations introduced in late 2007 that made illegal downloading difficult and that forced industry players to consolidate. A large number of new players, mainly state-owned or -controlled media enterprises, entered the market. For instance, the top search engine in China, Baidu, launched its video site Qiyi.com in early 2010 and now provides viewers with a variety of licensed and advertising-supported HD content, including free TV series, films, and variety shows. Current content partners include China Film Group, Huayi Brothers Media Group, Hunan Satellite TV, Beijing Satellite TV, and Zhejiang Satellite TV. The Baidu-backed site also has more than five hundred movies, both back-catalog and new releases, from Warner Brothers, Disney, Sony, Paramount, and Fox.

The rise in licensing costs creates flow-on impacts. Costs of overseas copyrighted content, especially Hollywood products, are high and are excessive for weakly financed online video sites. Because of cheaper licensing costs, East Asian pop culture has gained more presence on online video sites in China. Together with the effects of cultural proximity, the repositioning of East Asian content in the Chinese market contributes to an ethos of regional collaboration.

IMPLICATIONS FOR REGIONAL MARKETS

While the creative communities mentioned above represent the leading edge of innovation, they remain a mystery to government. Meanwhile, the five previous levels in the innovation timeline refer to formal mechanisms to initiate projects. Some of these strategies are simple; for example, find factory space, acquire workers, and look for outsourcing work. Hong Kong and Taiwan have played in this low-cost arena, setting up factories and using mainland China as a production base. The outputs of standardized production and imitation "take the low road,"[35] whereas the aim of collaboration, trade, and clustering is to travel a higher-value road. In much of the lower-value production-based activity, resources are committed to scale activities and price competition with a minimum of learning-by-doing.

In the third and fourth levels, actors seek out opportunities that have come with relaxation of media and cultural policy—for example, opportunities for cross-cultural alliances or for cross-platform production. This is where we see more and more pan-Asian networks, more and more creative migration from Hong Kong to Beijing and Shanghai in particular. Regional players are now scrambling to take up positions on the mainland, echoing Chua Beng-Huat's notion that the China market is now the driving force of East Asian pop culture.

China is becoming more open to investment; the national government is cautiously moving to encourage more collaboration, particularly in film and animation industries, while remaining vigilant about the need to build domestic capacity. Regional governments are actively encouraging pan-Asian collaborations. Inevitably, in the next decade regional producers, investors, and creative personnel will play a leading role in showing China's enterprises, officials, and people how to be creative.

NOTES

1. The full text of Hu Jintao's report at the Eighteenth Party Congress is available at http://news.xinhuanet.com/english/special/18cpcnc/2012-11/17/c_131981259_7.htm (accessed April 30, 2013).
2. For a discussion of Hengdian, see Michael Keane, *Created in China: The Great New Leap Forward* (London: Routledge 2007).
3. Cited in Chen Yingqun, "A city of talent for high-tech," *China Daily,* October 31, 2013.
4. For a discussion of industry classifications in China, see James McGregor, *No Ancient Wisdom, No Followers: The Challenges of Chinese Authoritarianism* (Westport, CT: Prospecta Press 2012).
5. Mao Zedong, "The Relationship between China and Foreign Countries 1956," cited in Li Lanqing, *Breaking Through: The Birth of China's Opening Up Policy*, trans. Ling Yuan and Zhang Siying (Oxford: Oxford University Press, 2009).
6. For a discussion, see Michael Curtin, *Playing to the World's Biggest Audience: The Globalization of Chinese Film and TV* (Berkeley: University of California Press, 2007).
7. Keane, *Created in China;* Michael Keane, *Creative Industries in China: Art, Design and Media* (London: Polity, 2013).
8. For a discussion of a "strong cultural nation" and its implications, see Yong Xiang, "2011–2015: Principles of national cultural strategy and cultural industries development in Mainland China," *International Journal of Cultural and Creative Industries* 1, no. 1 (2013): 74–80.
9. "Cultural soft power" is enshrined in Chinese government documents; it is an element of the broader matrix of soft power (*ruan shili*). The term *soft power* itself derives from the work of Joseph S. Nye Jr. In *Bound to Lead: The Changing Nature of American Power* (New York: Basic Books, 1990), Nye spoke primarily about soft power in terms of international diplomacy and foreign aid. Revisiting the concept almost two decades later in a foreword to a collection on Japanese "soft power," he conceded that he had underestimated the cultural impact of media industries. See Nye's foreword in W. Yasushi and D. McConnell, eds., *Soft Power Superpowers: Cultural and National Assets of Japan and the United States* (New York: East Gate, 2008).

10. See John A. Mathews, *Strategizing Disequilibrium and Profit* (Stanford, CA: Stanford University Press, 2006).
11. For a discussion, see Keane, *Creative Industries in China*.
12. For a discussion of this literature, see Shulin Gu and Bengt-Åke Lundvall, "China's innovation system and the move towards harmonious growth and endogenous innovation," *Innovation, Management, Policy and Practice* 8 (2006): 1–26.
13. The term *soft power competition* was coined by Chua Ben-Huat. See Chua Beng-Huat, *Structure, Audience and Soft Power in East Asian Culture* (Hong Kong: Hong Kong University Press, 2012).
14. Jici Wang, "Industrial clusters in China: The low road versus the high road in cluster development," in *Development on the Ground: Clusters, Networks and Regions in Emerging Economies*, ed. A. Scott and G. Garofoli (London: Routledge 2007), 145–64.
15. Joseph Schumpeter, *Business Cycles* (New York: McGraw Hill,1939); Joseph Schumpeter, *Capitalism, Socialism and Democracy* (London: George Allen and Unwin, 1939).
16. Wuwei Li, *Creativity Is Changing China*, ed. M. Keane; trans. M. Keane, H. Li, and M. Guo (London: Bloomsbury Academic, 2011), 36.
17. In the past, under the command economy presided over by Mao Zedong, cultural goods and services were produced for educational and political purposes.
18. See Chua Beng-Huat, *Structure, Audience and Soft Power*.
19. For a discussion, see Keane, *Created in China*.
20. Chua Beng-Huat refers to "soft power competition." See Chua Beng-Huat, *Structure, Audience and Soft Power*.
21. See Richard Caves, *Creative Industries: Contracts between Art and Commerce* (Cambridge, MA: Harvard University Press, 2000).
22. For a more comprehensive discussion of the six levels, see Keane, *Creative Industries in China*.
23. Johan Falk and Jonas Hagman, "Outsourcing to China? A Swedish perspective of business in the 21st century" (master's thesis, Lund University of Technology, 2002).
24. This is discussed by Judith Hollows. See Judith Hollows, "Trajectories of innovation and competitiveness: Hong Kong

firms and their China linkages," *Creativity and Innovation Management* 8, no. 1 (1999): 57–63.
25. Michael Keane, "It's All in a Game: Television Formats in the People's Republic of China," in *Rogue Flows: Trans-Asian Cultural Traffic,* ed. K. Iwabuchi, S. Muecke, and M. Thomas (Hong Kong: Hong Kong University Press 2004), 53–72.
26. Gillette Maris, "Copying, counterfeit and capitalism in contemporary China: Jingdezhen's porcelain industry," *Modern China* 36, no. 4: 367–403.
27. Brian Yecies, Aye-Gyung Shim, and Ben Goldsmith, "Digital Intermediary: Korean Transnational Cinema," *Media International Australia Incorporating Culture and Policy* 141 (2011): 137–145.
28. See Yuan Gao, "Lenovo's new secret weapon: Hollywood Star," *China Daily,* October 31, 2013.
29. For a discussion, see Michael Keane, Anthony Y. H. Fung, and Albert Moran, *New Television, Globalisation, and the East Asian Cultural Imagination* (Hong Kong: Hong Kong University Press, 2007); see also Dong-Hoo Lee, "From the Margins to the Middle Kingdom: Korean TV Drama's Role in Linking Local and Transnational Production," in *TV Drama in China,* ed. Y. Zhu, M. Keane, and R. Bai (Hong Kong: Hong Kong University Press, 2008), 187–200.
30. The "cultural exports deficit" refers to China's imbalance in terms of media and cultural exports to the world. Debate ensues about the validity of the data. Some reports, such as the United Nations Conference on Trade and Development's *Creative Economy Report,* claim that China is the world's leading cultural exporter, but the data are heavily skewed toward manufactured products such as furniture and appliances. See UNCTAD, *The Global Creative Economy Report 2010* (Geneva: UNCTAD, 2010).
31. See Keane 2011.
32. Alfred Marshall, *The Principles of Economics,* 8th ed. (1920; repr. Philadelphia: Porcupine, 1990).
33. Michael Porter, *The Competitive Advantage of Nations* (New York: Free Press, 1990).
34. For a discussion of copyright regulations impacting Chinese online media, see Elaine Jing Zhao and Michael Keane, "Between Formal and Informal: The Shakeout in China's Online Video

Industry," *Media, Culture and Society* 35, no. 6 6 [September 2013 vol. 35 no. 6]: 724–41.
35. Wang, "Industrial Clusters in China."

Chapter 2

Policy Transfer of Creative Industries for Economic Development: The Chinese Metropolitan Experience

Xuan-Olivia Jiang

The policy concept of creative industries was initiated in the United Kingdom in 1998, in an effort to change the stereotyped image of the nation and to accelerate economic restructuring. The emergence of the concept has attracted considerable attention and led to policy transfer. Policy transfer seldom results in carbon-copied policies in different places; policies usually change as they travel. Until recently, most studies on policy transfer have focused on advanced industrial countries, with few focused on accomplishments of other economies, such as China. Policy transfer in creative industries is little noticed. In this paper, Russell Prince's perspective of policy assemblage on policy transfer will be adopted to study how the objects of a transferred policy are constituted in different places, exemplified by the case of creative industries policy globally transferred. The objects of the policy transfer in this case include motives for initiating or adopting creative industries, the form of creative industries, statistical techniques, and detailed strategies. Also, problems will be discovered from the understanding and comparing of processes and results of the transfer of creative industries.

Sometimes learning things comes from secondhand or indirect experience due to their widespread existence and the low cost and high benefit of learning these things. Further, secondhand learning is usually through best practices replication or extrapolation.

The replication of best practices usually presents only simplified successful models without mentioning the details of experiences or failures during the processes. Therefore much information is missed and there is high risk to policy makers trying to replicate practices (Winter and Szulanski, 2001). Also, because of the different backgrounds of different places, replication is not realistic or workable. Extrapolation, a way of adaptively adopting successful models and inventing a more appropriate one under inspiration from the successful model, is contrarily favored. In this way the disadvantage of missing information could be avoided by utilizing the explained mechanism underlying the success. When the new model is established, it is expected that the mechanism leading to the success of the original model will also exist, which is the goal of extrapolation. To understand the mechanism in order to undertake extrapolation of secondhand experience, it is necessary to explain the mechanism underlying the successful model. On the one hand, the explained result, usually the conceptualized effects and performance of the successful practices, will be chosen; on the other hand, explanatory factors, such as the model design features, context, and participants' interaction, need to be explored to interpret the mechanism driving the success. With the interpreted mechanism, experiential learners will be able to design a more applicable and responsive model to fit the particular situation and to solve the particular problem (Bardach, 2004, 2003; Barzelay, 2007; Caulkins, 2002; Christophers, 2003; Elster, 1998; Mackie, 1993). In the field of policy making, extrapolation-based secondhand experiential learning provides a good analytical and explorative regime for cross-place policy learning—that is, policy transfer.

The process of policy transfer between places has attracted growing interest among analysts in recent years (Evans, 2004; McCann, 2008; Peck and Nik, 2001; Stone, 2004; Ward 2006), even though it has existed since organized government was first established. In recent years, interest has been prompted by a growth in transfers in an increasingly interconnected world. This refers to the manner in which policies developed in a particular time and place come to influence the development of a similar policy in another time and place (Dolowitz and Marsh 1996). It thus implies the internationalization and transnationalization of policy regimes and emphasizes the people, places, and moments through which this situation eventuates. The results of policy transfer are seldom carbon-copied. On the contrary, inspired by the mechanism underlying transferred policy,

policy makers at different localities usually change policies as they travel to fit local social, economic, political, and cultural contexts (Peck and Nik, 2001; Phelps et al., 2007).

Today, in a time of globalization, policy transfer has become an important policy-making approach, solving domestic problems and decreasing policy learning costs. However, the process of policy transfer is not always smooth or effective. A variety of factors, such as different ideological, political, and administrative issues; local resources; unclear intention and interaction; and complex policy design features, may all hinder the successful transfer of the model policy from the place of origin and even result in failure. Although extrapolation-based cross-place policy learning is a good regime for policy makers where the model policy is transferred to, in many scenarios, it's not easy to undertake extrapolation due to insufficient information about the context, design features, and particularly participants' interactions behind the model policy making. Therefore the mechanism leading to success is unclear. Due to the increasing importance of policy transfer, blurry processes and mechanisms of model policy making, and uncertainty regarding results, it is necessary to undertake further empirical and theoretical studies on policy transfer.

The emergence of creative industries, accompanied by globalization and the knowledge economy as well as by the growing importance of cultural industries (Flew, 2002), was formally initiated in the United Kingdom in 1997 as a national economic development and image-changing strategy (DCMS, 1998). Since then, it has attracted much attention from academics, and policy makers have widely promoted it in countries and cities around the world. Until recently, most studies on policy transfer have focused on advanced industrial countries, with few focused on accomplishments of other economies, including China. This paper adds to the discussion by showing how the objects of a transferred policy are constituted in different places, such as China, and what problems might ensue from this constitution.

THE DEVELOPMENT OF CREATIVE INDUSTRIES AS A GLOBAL PHENOMENON

The emergence of creative industries is accompanied by globalization and the knowledge economy, as well as by the growing importance of cultural industries (Flew, 2002). The concept of creative industries was brought forward by the newly established Creative Industries Task Force in the United Kingdom in 1997 soon after the new Tony Blair government took over political control (DCMS, 1998). Promised in the "Cool Britannia" campaign of the New Labor Party during the election, the concept was a strategy to attract outside attention and to change the traditional national economic structure and the stereotyped image of old manufacturing industries.

To further promote the new concept, the UK Department for Culture, Media and Sport published the *Creative Industries Mapping Document* (CIMD), which defined creative industries as "those industries which have their origin in individual creativity, skill and talent and which have a potential for wealth and job creation through the generation and exploitation of intellectual property" (DCMS, 2001: 4). The document also delineated industries, including advertising, architecture, arts and antiques, crafts, design, designer fashion, film and video, interactive leisure software, music, the performing arts, publishing, software and computer services, and television and radio (DCMS, 2001). It also technically mapped and evaluated economic contributions of these industries in terms of economic indexes including revenue, exports, and employment, indicating the importance of creativity to the British economy (Christophers, 2007). Although not stated in the CIMD, the department made accordant policies and strategies to help the development of creative industries. The goals of the policies were to incubate self-sustained markets and to establish creative milieu. In general, the policies included providing fiscal initiatives and other support for enterprises (especially SMEs), strengthening human resource development systems, building networks to strengthen that support, protecting intellectual property rights, nurturing the creative lifestyle and environments, branding creative industries, and incubating creative clusters.

Optimistic economic estimates of creative industries in the United Kingdom affirmed the value of the policy concept of creative industries. Now the concept has been promoted on a global scale, and a large number of countries, such as the United States, Australia, New Zealand, Germany, Sweden, Korea, Japan, Singapore, Fiji, and China have adopted this idea. The trend continues.

Policies change as they travel due to different local contexts and conditions, including culture, socioeconomy, politics, ideology, history, and resources. Policy transfer involves tactics designed for particular local conditions and problems, making extralocal policy programs more responsive and applicable in local contexts (Prince, 2010). When the policy program of creative industries is adopted, the vague definition of the concept gives localities much freedom to design programs to fit their local circumstances and policy implications. Thus a pattern of global variation appears, and the idea of creative industries varies in terms of definition and industrial components among nations, between a nation and its urban centers, and among urban centers in the same nation. Sometimes the term *creative industries* is used interchangeably with the terms *cultural industries, copyright industries,* and *cultural and creative industries*. For example, the "creative industries" in Hong Kong fit the UK definition to capture the local importance of IT and digital media (Hong Kong University, 2003). The "copyright industries" in the United States and Australia emphasize legal registration of intellectual property as an increasingly important mechanism to protect creative products and activities (Siwek, 2004). The "cultural industries" in Denmark emphasize culture to prevent cultural assimilation from globalization (BOP Consulting, 2010). The "cultural and creative industries" in China, officially brought forward through the document *National Planning Program for Cultural Development during the Eleventh Five-Year Period,* issued by the central government, include eight major categories reflecting the movement of recent economic restructuring toward industrial sectors with high benefits and technical innovation and assisting national cultural reform toward marketization of cultural affairs with government supervision of the creative industries market to control the ideology of cultural products. Among urban centers in the same nation, Bristol adopted the definition established by the national UK government, which is different from that used by London (Bristol City Council, 2011). In China the sectoral boundary of "cultural and creative industries" in Beijing is close to that of "cultural industries" defined by the China National Bureau of Statistics in 2004 to fit Beijing's position as the national cultural capital. The definition differs from that for "creative industries" in Guangzhou, reflecting its long-term strength in creative manufacturing and industrial sectors. Despite variations in different localities, the industrial components of creative industries are usually a mix of art and cultural industries of the preindustrialization period, creative manufacturing sectors of the industrialization period, and software and digital sectors of the postindustrial period (Evans, 2009).

To successfully develop creative industries, governments issue policies and strategies at different places. These policies and strategies are generally similar from place to place. The policies and strategies made by the UK government work as how-to manuals to guide the development of creative industries in other countries and regions. The interventions involve supporting the incubation of creative clusters and establishing mature markets through providing various combinations of infrastructure, facilities, financial assistance, investment opportunities, information services, intellectual property protection, business development support, and market expansion support to enterprises; strengthening the labor pool; encouraging cross-department collaboration and public–private partnership; branding creative industries through events; and cultivating culture as a lifestyle. The policies and strategies are not industry-specific but are for general enterprise development.

The Policy Transfer of Creative Industries

Policy transfer can be divided into two successive steps (see Figure 2.1). First, policy transfer involves policy tailoring, by which policy objects become workable in the place the policy is being transferred to. This process of decontextualization, delocalization, and universalization gives the model policy objects a neutral global form. The second step is policy assemblage, making the neutral global form articulated in a particular place. This is the process of recontextualization and localization of the neutral global form; the assemblage of texts, actors, agencies, institutions, and networks at the place of adoption (Collier and Ong, 2005; Prince, 2010). However, the process of policy tailoring may take away many details of model policy making at the place of origin. As noted at the beginning, extrapolation-based secondhand experiential learning requires an understanding of the mechanism leading to the success of the model policy, so it is necessary to figure out the context, design features, and interactions behind the model policy making. The process of policy tailoring can make it hard to figure out these critical explanatory factors, making it impossible to correctly determine the mechanism.

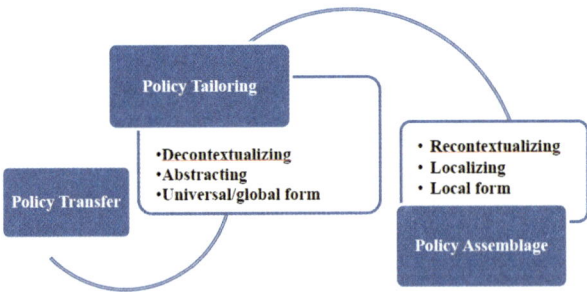

Figure 2.1 The anatomized process of policy transfer

The creative industries also go through the two-stage process. As noted earlier, in the United Kingdom, four efforts taken by the government encouraged global policy transfer: defining the concept, identifying the industrial components, mapping and measuring the economic contribution, and providing details and strategies for development, with measured economic contribution making creative industries noticeable and efforts for promotion reasonable and with the other three efforts making creative industries respectively knowable, identifiable, and workable in other places. These efforts also decontextualize the development of creative industries in a global form composed of the must-have conceptual definition, sectoral boundary, and policies and strategies. Here, measured economic performance is seen as the explained success of the creative industries in the United Kingdom, and the other three are seen as the model design feature. In particular, the policies and strategies are seen as the mechanism that can lead to success. In most places of adoption, policy makers believe that to develop creative industries, the three design features are very necessary at the beginning. Therefore, the process of global policy formation is seen as technical, neutral, and decontextualized to make creative industries more workable to the particularity of place during later assemblage. At the second stage of policy transfer—policy assemblage—to make the global policy program of creative industries flexible, responsive, and operable at localities, each place conceptually and sectorally defines creative industries according to their particularities (Prince, 2010). Policies and strategies, seen as mechanisms leading to success, basically remain the same.

Policies change as they travel due to different local contexts. The second stage of policy transfer is the process of policy assemblage, which makes a particular global form articulated in a particular place. The difference between places means that policy assemblage involves policy makers' tactics designed around particular problems and making extralocal policy programs applicable in local contexts (Prince, 2010). Looking back to the time when the concept of creative industries was born, and with more information about why and how it was born, we find that the intention was far from neutral but very political.

When the concept of creative industries was initiated in the United Kingdom, when the New Labor Party took political power, the intention was political rather than economic. As noted earlier, the Creative Industries Task Force was established to develop creative industries under the administration and management of DCMS (which used to be named the Department of National Heritage). The new head of DCMS, Chris

Smith—now working in the cabinet—had long been associated with New Labor's cultural policies (Smith, 1998). The strong personal enthusiasm of Smith and the long-term ignorance of the UK government regarding arts and culture prompted Smith to put much priority and emphasis on cultural policy and cultural industries in particular, which marked a new status for the cultural agenda. A DCMS series firmly established the cultural industries as a legitimate object of policy (DCMS, 1998). Backed by optimistic and convincing economic contributions from the thirteen subsectors, the cultural industries, previously ignored or lumped together with "the arts," were officially placed at the heart of local and regional cultural and economic strategies after long neglect (DCMS, 2000, 2004). In fact, among the 5.4 percent contribution of creativity to the British economy in 1997, 1.7 percent came from the software, computer games, and electronic publishing industries and 1.1 percent came from the publishing industry, with arts and cultural activities contributing little. However, under the umbrella of "creative industries," the cultural industries were also seen as critical. The smallest and newest department, DCMS became bold and began talking to the very large and well-established Department of Trade and Industry—something that would have been unthinkable earlier (Kong et al., 2006). Meanwhile, the emergence of the concept of creative industries ideologically represented a new political idea to change the image of "Old Labor" represented by Margaret Thatcher (Blair, 1998; Gibson, 2001). The industrial components, including industrial sectors such as software and heritage, with creativity as the core, transcended the traditional division between "high" and "popular" culture (Hughson and Inglis, 2001). This meant that all cultural forms could be celebrated for their contribution to British identity (Smith, 1998). This was a reassertion of the forces of social justice against those of hard-headed economics (Redhead, 2004). Apparently, the initiation of creative industries was full of political color, delivering more political and personal intention. No wonder the global pattern of development of creative industries turns out to be widely varied.

Policy Assemblage for the Creative Industries in China

Creative industries matter to advanced industrial countries or regions for improved competitiveness, relying less on price than on creativity. The nations are less likely to lose out to price-led competition, a force that has caused many manufacturing and service jobs to be outsourced to emerging economies. In recent years, emerging economies have been eager to reposition themselves to benefit from economic globalization, move away from price-led competition, reduce the gap with advanced economies, identify new sources of competitive advantage, and acquire economic independency from the developed world (Liefner et al., 2006). In this context, developing creative industries, a critical way of shifting toward higher-value-added industries and getting technical independence, has become a desirable approach for emerging economies. This is clearly the case for China.

In China in 2006, the concept of "cultural and creative industries" was officially brought forward at the national level through the *National Planning Program for Cultural Development during the Eleventh Five-Year Period*. The release of the document symbolized acknowledgment of and emphasis on cultural and creative industries by the national government of China. However, in the central government, no clear conceptual or sectoral definition was determined, and no detailed policy strategy—other than general guidance to encourage cities to play a leading role—was made. In response to such deficiency, the China Academy of Social Sciences (CASS), a nongovernmental, national-level research institute, set up a standard sectoral boundary for creative industries in China in 2007. It includes eight major categories, twenty-one medium-level categories, and seventy-five smaller categories, using the classification of the creative industries in the United Kingdom as a benchmark. Like CIMD bluebooks, the *Annual Report on the Creative Industries in China* has also been issued, providing economic estimates of industries periodically. It was estimated that in 2005, among business revenues from all eight subsectors, the consulting industry generated 23 percent and the telecommunication and software industry contributed 48.1 percent.

However, the emergence of cultural and creative industries in China was not only for economic restructuring and innovation but also in the context of recent culture reform. First, using the term *cultural and creative industries* rather than *creative industries* helps highlight the current emphasis on the role of culture in economic development and image building. With reforms in economic sectors accelerating in the wake of the Fourteenth Congress of the Chinese Communist Party, which ushered in the socialist market economy in 1993, further resolutions soon followed. In particular, a 1996 document attempted to separate institutional functions from business enterprise functions and to separate political institutions from public services in order to conduct business operations as autonomous legal entities. In 1996 the Guangzhou Press Group was formed. Meanwhile, an important reform began: separating production and broadcasting in the broadcast media. In October 2000 the phrase *enhancing cultural industry development* appeared in public documents. In August 2001 the Chinese government issued a document affirming the idea of allowing cultural institutions to form into business groups. In June 2003, at a conference on the trial operation of reforms in cultural institutions, the practice of reform from institution to industry was adopted, and this led to wider reforms in the organization of cultural institutions. Second, since a number of creative industry sectors, such as publishing and mass media, are ideology-relevant, highlighting culture will help control the production and distribution of some cultural products. The essence of the reform of the cultural system is setting up a new cultural system to fulfill the demand of economic and cultural development. The core spirit of the new cultural system is that the government is not the provider of cultural production. The duty of the government is to guarantee that cultural production and services are provided efficiently. For those that the market can provide, it should be allowed to do so. For those that the market cannot support, owing to market failure, the government should seek ways to ensure compensation (Zhang, 2006). This leaves much space for government to intervene in the cultural market as much as it wants.

Policy transfer between states has attracted interest among analysts in recent years. The literature conceives transfer as relatively discrete: policy formulation begins with transfer processes, which are then replaced by domestic processes during implementation (Prince, 2010). As noted, no detailed policy strategy, other than general guidance to encourage cities to play a leading role, was made, so in China cities are the main entities to develop creative industries. Now the policy concept of creative industries has been transferred all over China. Governments at the municipal level have explicitly promoted creative industries through the issuing of formal government documents. Until 2010, more than twenty of the forty-one major municipalities with a population over 1 million (including Beijing, Shanghai, Tianjin, and Chongqing) officially promoted creative industries. Six creative-cities regions have been formed: the Capital Region, the Yangtze River Region, the Pearl River Region, the Southwestern Region, the Western Region, and the Central Region.

Among those cities promoting creative industries, a large number stopped at adopting the term and issuing government documents. A few, such as Beijing, Shanghai, and Guangzhou, have undergone the process of policy assemblage.

POLICY ASSEMBLAGE FOR THE CREATIVE INDUSTRIES IN SHANGHAI, BEIJING, AND GUANGZHOU

Shanghai and Beijing are by far the most advanced in terms of economic performance of the creative industries. The economic performance of creative industries in Guangzhou ranks third in terms of the number of enterprises, employment, and business revenue, measured using the CASS sectoral boundary.

Shanghai and Beijing, the national centers of China, and Guangzhou, a regional economic center, all have advanced economic structures, various cultural resources, strong academic institutions, and good business environments. They have played leading roles in China's economic growth in recent decades. Shanghai and Beijing have successfully built images as major cities in the global system. Guangzhou has also been eager to improve its overall competitiveness and to emerge onto the global stage. With different visions, creative industries—globally acknowledged tools for urban development—have been adopted by the governments of these three cities. The municipal governments in Shanghai and Guangzhou have formally promoted creative industries since 2005, and Beijing has done so since 2006. As policy-transferred cities, all have used a four-legged-stool approach to accept creative industries as a global form, except that Guangzhou has not provided a definition.

The municipal government of Beijing categorizes the cultural and creative industries into nine general groups, twenty-seven medium-level groups, and eighty-eight smaller groups. The municipal government of Shanghai categorizes the creative industries into five general groups, thirty-eight medium-level groups, and fifty-five detailed groups. In Guangzhou, no official sectoral boundary has been made for creative industries, and the Guangzhou Association for Social Sciences (GASS), a major research institute, categorizes them into five general groups, twenty-three medium-level groups, and seventy-two smaller groups. The sectoral boundary of creative industries in Shanghai is the narrowest, representing what the government would like to focus on moving forward. Defined as "interrelated industries providing cultural experiences," the cultural and creative industries in Beijing are closest to the term *cultural industries* as defined by the China National Bureau of Statistics in 2004. They uniquely involve the value chain of culture-relevant products, which fits the city's position as the cultural capital of China. The creative industries in Guangzhou uniquely involve the manufacturing industries for textile, fur, leather, and feather products, reflecting the city's long-term strength in these sectors. The locally defined creative industries reflect local economic backgrounds or imply the direction of economic restructuring.

Each city has a city-level research institute providing reports for the local creative industries. The Guangzhou Association of Social Sciences issues the *Report of the Creative Industries in Guangzhou;* the Beijing Association of Social Sciences issues the *Report of the Cultural and Creative Industries in Beijing;* and the Shanghai Association of Social Sciences issues the *Report of the Creative Industries in Shanghai*. Based on CASS definitions and data from the 2004 economic census, it turns out that creative industries make sizable contributions to the local economies of these three cities. The share of creative industries in overall employment in the local economy ranges from 8.8 percent in Shanghai to 23 percent in Guangzhou. The share of creative-industry business revenues in the overall local economy ranges from 6.1 percent in Guangzhou to 8.8 percent in Beijing.

In all three cities, various policies for supporting creative industries in general or in particular sectors (for example, software and animation industries) have been issued. Policies from each of these three municipalities are similar and have the goal of completing the value chain and establishing mature markets for creative industries. The policies generally include providing special funds; training, awarding, and attracting staff; fostering the growth of enterprises; branding creative industries; encouraging R&D and originality; and building infrastructure such as creative industries parks. Policies in China tend to focus on the supply side of the market, with less attention to stimulating or nurturing consumer demand. Also, in Shanghai government documents explicitly emphasize the importance of SMEs in creative industries and particularly require the provision of financial assistance to them. In contrast, the municipal government of Guangzhou tends to support the growth and development of large-scale enterprises in creative industries.

Comparing Chinese Experience with the Global Situation

The policy assemblage process helps reveal the similarities and disparities between Chinese experience and that in a wider context. It has been found that the current situation of creative industries in China reflects that occurring around the world in many aspects, including:

Creative industries make a positive contribution to economic and urban development.

- The strategy of developing creative industries for economic and urban development has been widely diffusing around the world and within China.

- Creative industries vary across locations and scales. Different understandings, preferences, study intentions, and policy implications all contribute to the variation, which attests to the flexibility of the notion to meet various needs.

- Policies for developing creative industries are similar.

- Aiming to establish a mature market, governments foster enterprises, incubate clusters, attract and retain workforces, brand products, build partnerships, and so on. At the same time, China also has several specialties. In terms of policy making, policies in China are generally similar to those in developed countries. However, policies in Shanghai, Beijing, and Guangzhou focus on only the supply side of the market and neglect the demand side. This is different from policies in a wider context, in which both sides of the market are emphasized. Also, the Guangzhou government particularly supports large enterprises in creative industries, contrary to the global trend, which emphasizes support for SMEs.

Conservative Policies for New Economy

In terms of policy making, the government in Guangzhou and the developed economies tend to be conservative. Policy makers use new bottles for old wine and ignore the potential needs of new industries. Although vague definitions result in fuzzy sectoral boundaries for the creative industries, industrial components in different places are still from three general categories, according to local strengths and needs: post-Fordism industrial sectors (for example, animation, software, and digital technology), Fordism industrial sectors (for example, manufacturing), and pre-Fordism industrial sectors (for example, arts and tourism) (Evans, 2009). Even to promote new economies represented by post-Fordism industrial sectors, traditional economic- and enterprise-development approaches are still used, such as providing infrastructure and human capital and branding creative products.

Also, when policy makers face industrial sectors with wide discrepancy, such as culture and animation industries, they use a one-model-for-all strategy to promote them. It is assumed that policy needs of industrial sectors with vast differences are the same. For example, it is assumed that cluster incubation, the major strategy for developing creative industries or cities, is the common need of all industries, ignoring the particular characteristics and need of each subsector. Therefore, policies for developing creative industries are seen as enterprise-oriented rather than industry-oriented (Evans, 2009).

However, if current policies are criticized as lacking imagination and innovation, do more appropriate strategies exist? What will the new strategies be? Will empirical studies prove that an imagined innovative policy can be more effective in promoting creative industries or creative cities?

LARGE-SCALE ENTERPRISES OR SMES

The size of enterprises to be promoted is also a critical issue. Compared with large-scale enterprises, SMEs are pioneers and major participants in new areas of business, partly because of lower entry costs. Also, interaction between SMEs and the flexibility of SMEs tend to bring business vitality to the market (Honjo and Harada, 2006). A decrease in the number of SMEs may cause a decline in the prosperity, diversity, and sustainability of the market, so a lack of funding resources, R&D, human capital, and information prompts governments to issue policies to support them. However, the government of Guangzhou presents a different model by paying more attention to large enterprises, even though SMEs are still the main business form in the creative market of the city. In Guangzhou, the policies of the municipal government are not for the full-blown creative industries but for two subsectors—the software and animation industries. In the city, SMEs are the predominant business form of the software industry. Enterprises with assets less than 5 million yuan (about US$700,000) account for 88.1 percent of Guangzhou's software firms, while those with assets of less than 500,000 yuan (about US$70,000) account for 58.3 percent of software firms. Since the development of the creative industries has been fostered for a relatively short time, it is hard to observe the results of the policies. The reason for focusing on large-scale enterprises when SMEs are the main business form in the creative industries in the city merits further exploration.

However, the global trend of supporting SMEs in creative industries is also debatable, since the growth of SMEs seems to be overemphasized, with the needs of large enterprises being neglected. Although the SME is the major business form in the creative market in terms of numbers, large creative enterprises dominate the market in terms of economic impact, therefore drawing more support from outsiders (Evans, 2009). An empirical study shows that in developed European cities, SMEs, which account for about 90 percent of the market in terms of numbers, did not have much economic impact. They play an important role in the market in a grassroots, bottom-up way, but the low average number of employees in SMEs make them fail to generate satisfactory employment and value-added rates (Foord 2009). Therefore, whether policies should be more inclusive to pay attention to enterprises of all sizes merits further deliberation.

CONCLUDING OBSERVATIONS

Policy transfer study provides a fresh geographical perspective on uneven processes of restructuring in the global political economy (Peck and Nik, 2001). The fast transfer of the concept of creative industries in policy over a relatively short time can be attributed to the successful transfer of global policy formation from the United Kingdom to a large number of other contexts and to successful policy assemblage in different localities (Prince, 2010). Dividing policy transfer into global form making and policy assemblage is powerful as a way to abstract factors that are universal for policy transfer and to further examine what special tactics are utilized to make a policy applicable and workable at localities. The process and result of the second step also reveal many similarities and disparities between efforts in different localities, making comparative studies worthwhile in discovering critical problems and raising debates. Factors making a policy transfer start will not ensure success of the policy transfer. Anatomizing policy transfer into two stages will help determine sufficient conditions for the success of creative industries as thoroughly as possible.

BIBLIOGRAPHY

Bardach, Eugene. 2003. "Creating Compendia of Best Practice." *Journal of Policy Analysis and Management* 22: 661–88.

———. 2004. "The Extrapolation Problem: How Can We Learn from the Experience of Others?" *Journal of Policy Analysis and Management* 23 (2): 205–20.

Barzelay, Michael. 2007. "Learning from Second-Hand Experience: Methodology for Extrapolation-Oriented Case Research." *Governance: An International Journal of Policy, Administration, and Institutions* 20 (3): 521–43.

Blair, Tony. 1998. *The Third Way: New Politics for a New Century.* London: Fabian Society.

BOP Consulting. 2010. *Mapping the Creative Industries: A Toolkit.* London: British Council.

Bristol City Council. 2011. *Creative Industries.* Business Match, Bristol City Council, European Union, http://www.businessmatch.org.uk/634.asp [accessed August 1, 2012].

Castells, Manuel. 1996. *The Rise of the Network Society.* Oxford: Blackwell.

Caulkins, Jonathan. 2002. "Using Models That Incorporate Uncertainty." *Journal of Policy Analysis and Management* 21 (2): 486–91.

Christopher, Gail. 2003. "Innovative Government Practice: Considerations for Policy Analysts and Practitioners." *Journal of Policy Analysis and Management* 22 (4): 683–88.

Christophers, Brett. 2007. "Enframing Creativity: Power, Geographical Knowledges and the Media Economy." *Transactions of the Institute of British Geographers* 32 (2): 235–47.

Collier, Stephen, and Aihwa Ong. 2005. "Global Assemblages, Anthropological Problems." In *Global Assemblages: Technology, Politics, and Ethics as Anthropological Problems*, edited by Stephen J. Collier and Aihwa Ong, pp. 3–21, Malden, MA: Blackwell.

Department for Culture, Media and Sport (DCMS). 1998. *Creative Industries Mapping Document.* London: Department for Culture, Media and Sport.

———. 2000. *Creative Industries: The Regional Dimension.* London: Department for Culture, Media and Sport.

———. 2001. *Creative Industries Mapping Document.* London: Department for Culture, Media and Sport.

———. 2004. *The DCMS Evidence Toolkit*. London: Department for Culture, Media and Sport.

Dolowitz, David, and David Marsh. 1996. "Who Learns What from Whom: A Review of the Policy Transfer Literature." *Political Studies* 44 (2): 343–57.

Elster, Jon. 1998. "A Plea for Mechanisms." In *Social Mechanisms: An Analytical Approach to Social Theory*, edited by Peter Hedstrom and Richard Swedberg, pp. 45–73. New York: Cambridge University Press.

Evans, Graeme. 2009. "Creative Cities, Creative Spaces and Urban Policy." *Urban Studies* 46 (5–6): 1003–40.

Evans, Mark. 2001. *Policy Transfer in Global Perspective*. Ashgate, UK: Hants.

Gibson, Lisanne. 2001. *The Uses of Art: Constructing Australian Identities*. Brisbane: University of Queensland Press.

Flew, Terry. 2002. "Beyond Ad Hocery: Defining Creative Industries." In *The Second International Conference on Cultural Policy Research*, conference proceedings. Wellington.

Flew, Terry, and Stuart Cunningham. 2010. "Creative Industries after the First Decade of Debate." *Information Society* 26 (2): 113–23.

Foord, Jo. 2009. "Strategies for Creative Industries: An International Review." *Creative Industries Journal* 1 (2): 91–113.

Hong Kong University. 2003. *Hong Kong Creative Industries Baseline Report*. Hong Kong: Cultural Policy Research Center, Hong Kong University.

Honjo, Yuji, and Nobuy Harada. 2006. "SME Policy, Financial Structure and Firm Growth: Evidence from Japan." *Small Business Economics* 27: 289–300.

Hughson, John, and David Inglis. 2001. "'Creative Industries' and the Arts in Britain: Towards a 'Third Way' in Cultural Policy?" *International Journal of Cultural Policy* 7: 457–78.

Kanai, Miguel, and Iliana Ortega-AlcaZar. 2009. "The Prospects for Progressive Culture-Led Urban Regeneration in Latin America: Cases from Mexico City and Buenos Aires." *International Journal of Urban and Regional Research* 33 (2): 483–501.

Kong, Lily, Chris Gibson, May Khoo, and Louise Semple. 2006. "Knowledges of the Creative Economy: Towards a Relational Geography of Diffusion and Adaptation in Asia." *Asia Pacific Viewpoint* 47 (2): 173–94.

Mackie, John. "Causes and Conditions." 1993. In *Causation*, edited by E. Sosa and M. Tooley, pp. 33–55. New York: Oxford University Press.

Markusen, Ann, Gregory Wassall, Douglas DeNatale, and Randy Cohen. 2008. "Defining the Creative Economy: Industry and Occupational Approaches." *Economic Development Quarterly* 22 (1): 24–45.

McCann, Eugene. 2008. "Expertise, Truth, and Urban Policy Mobilities: Global Circuits of Knowledge in the Development of Vancouver, Canada's 'Four Pillar' Drug Strategy." *Environment and Planning A* 40: 885–904.

Peck, Jamie. 2009. "Recreative City: Amsterdam, Vehicular Ideas and the Adaptive Spaces of Creativity Policy." *International Journal of Urban and Regional Research* 36 (3): 462–85.

Peck, Jamie, and Theodore Nik. 2001. "Exporting Workfare/Importing Welfare-to-Work: Exploring the Politics of Third Way Policy Transfer." *Political Geography* 20: 427–460.

Phelps, Nicholas, Marcus Power, and Roseline Wanjiru. 2007. "Learning to Compete: Communities of Investment Promotion Practice in the Spread of Global Neoliberalism." In *Neoliberalization: States, Networks, Peoples,* edited by K. England and K. Ward, pp. 83–109. Blackwell: Oxford.

Prince, Russell. 2009. "Globalizing the Creative Industries Concept: Traveling Policy and Transnational Policy Communities." *Journal of Arts Management, Law, and Society* 40 (2): 119–39.

–––. 2010. "Policy Transfer as Policy Assemblage: Making Policy for the Creative Industries in New Zealand." *Environment and Planning A* 42 (1): 169–86.

Redhead, Steve. 2004. "Creative Modernity: The New Cultural State." *Media International Australia* 112: 9–27.

Siwek, Steve. 2004. "The Measurement of 'Copyright' Industries: The US Experience." *Review of Economic Research on Copyright Issues* 1 (1): 17–25.

Smith, Chris. 1998. *Creative Britain.* London: Faber and Faber.

Stone, Diane. 2004. "Transfer Agents and Global Networks in the 'Transnationalization' of Policy." *Journal of European Public Policy* 11 (3): 545–66.

Wang, Jing. 2004. "The Global Reach of a New Discourse: How Far Can 'Creative Industries' Travel?" *International Journal of Cultural Studies* 1: 9–19.

Ward, Kevin. 2006. "Policies in Motion, Urban Management and State Restructuring: The Trans-local Expansion of Business Improvement Districts." *International Journal of Urban and Regional Research* 30: 54–75.

White, Andrew. 2009. "A Grey Literature Review of the UK Department for Culture, Media and Sport's Creative Industries Economic Estimates and Creative Economy Research Programme." *Cultural Trends* 18 (4): 337–43.

Winter, Sidney, and Gabriel Szulanski. 2001. "Replication as Strategy." *Organization Science* 12 (6): 730–43.

Zhang, J. 2006. *Report on the Development of Creative Industries in China*. Beijing: China's Economy Publishing.

———. 2007. *Report on the Development of Creative Industries in China*. Beijing: China's Economy Publishing.

Chapter 3

Cultural Revitalization and Pedestrianization in Chengdu and Kunming: Lanes and Streets

Anthony K. C. Ip

With Western development thrusts in China in the 2000s, several major cities have embarked on cultural revitalization and pedestrianization in their urban cores. In the historic cities of Chengdu and Kunming, the conventional roles of enclave, haven, and outpost have been transformed into one of new regional economic-technological hub and Southeast Asian gateway. The development resources cut across innovations in everything from high-level industrialization, education, mineral and precious-metal extraction, agriculture, and biotechnology to cultural industries and urban lifestyles. Their synergistic nature is transforming and drawing new development paradigms and investments, with many capitalizing on the rich Han and minority cultures. This paper identifies profiles of the Wide and Narrow Lanes district and other old streets and analyzes their urban design and planning aspects within certain social, economic, technological, and policy contexts as foundations for further research and development.

INTRODUCTION TO WESTERN CHINA: FROM AFFLUENT COAST TO THE WILD WILD WEST

With the success of the open-door policy and economic reform in the past three-plus decades (1978–2013), China has maintained a centrally planned development policy skewing toward its geographic west. In 2012, with a national GDP growth rate of over 8 percent, the western region for the first time exceeded the affluent coastal cities. The western region includes six provinces—Gansu, Guizhou, Qinghai, Shaanxi, Sichuan, and Yunnan—with Chongqing now a municipality reporting directly to the central government. Their combined population is about 260 million. However, coupled with the high growth are social issues such as transformation from a predominantly agricultural to an urban economy; labor, knowledge, and skill needs; infrastructural support; housing; health care; and related lifestyle issues, particularly for the new diaspora. There have been quantum advances in information, communication, and technology in the cities, where many areas are physically deteriorating and can't support high growth and modern living. To understand the scale of such urban challenges, consider that there are more than one hundred towns and cities with populations over 1 million to be revitalized in the next twenty years. Although isolated from the coasts and central plains, Chengdu and Kunming have received much attention since the mid-2000s. Many might ask, "Will the West be the last mass market in China?" In this study, Chengdu and Kunming are selected for their commonalities and differences amid the context of urban development as an "enforcer" for nation building. Meanwhile, economic development has imposed tremendous pressures on the urban core, in particular in areas with high cultural heritage value.

Figure 3.1 Wide and Narrow Lanes in the past

Rising property and land values, rising rents, and the proliferation of cars are some of the key issues. On the other hand, Chengdu and Kunming inherently possess some unique cultural resources—for example, the mixed Han (漢), Manchu–Mongol (滿蒙), Basu (巴蜀), Hani (哈尼), and other minorities and their changing urban forms offer some good lessons. In brief, this paper provides background and policy on economic, social, and technological forces in Chengdu and Kunming, with specific areas demarcated for analysis of cultural revitalization and urban forms.

CHENGDU: REGIONAL AND URBAN SETTING

Geographically, Chengdu is an enclave of a plateau in southwestern China. To its northeast is Wuhan (武漢) and northwest is Tsinghai (青海). To the east is a series of mountains and gorges to the middle Yangtze, with Guizhou (貴州) on the southwest. Chongqing (重慶) is an autonomous city to the southeast. Aside from the symbolic panda, due to panda bears' habitation there, Chengdu has recently been actively revitalizing its urban centers and nearby towns. Among them is an area called Wide and Narrow Lanes (寬窄巷子), which housed barracks for Eight Flag reserve regiments dating to the late Qing Dynasty. It is located in the Qingyang district (青羊區) adjacent to the old city center. This was a trading center for silk, tea, porcelain, and agricultural products such as pepper. Today Chengdu is well linked by conventional rail, bus, and highways with Kunming and the Southwest in general. Residents can also reach Chongqing via high-speed rail in an hour. Historically, Chengdu was often romanticized as the Country of Heaven (天府之國) or a land of abundance. The Min River (泯江) to the west of Chengdu flows from the northern mountains to the south before turning into a major tributary of the middle Yangtze River. At the upper Min River, Mount Qingcheng (青晨山), a key Taoist retreat, rises to an elevation of 1,600 meters.

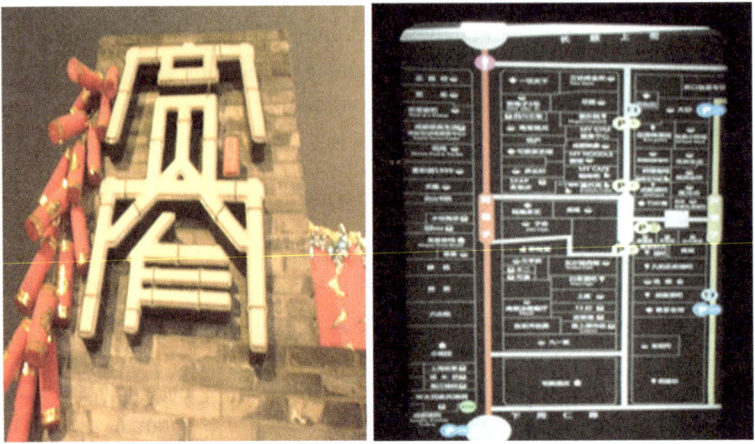

Figures 3.2.1 & 3.2.2 Wide and Narrow Lanes today

CHANGING CHENGDU'S URBAN CORE: FROM DEFENSE ENCLAVES AND BARRACKS TO A TRIPLE-LANE COMMUNITY

Chengdu is comprised of old and new physical forms, with rich historical and cultural content. The old city walls were mainly for defense, constructed of mud, stone, and brick over many centuries, but regrettably few relics still stand in the city center. However, the Basu lifestyles have continued in small and medium-sized vernacular low-rise residential houses, as well as temples, teahouses, historic villages, and towns. While landmarks are often accentuated with religious significance, flying cornices, and other symbolic and decorative motifs, there is a combination of wood, brick, and stone structures. Examples are Daci Monastery (大慈寺), Wenxu Court (文殊院), Wide and Narrow Lanes (宽窄巷子), and the Wuhou Shrine (Temple of Marqui Wu 武候祠). Before the 1900s, travel was by horse, cart, and foot along dirt roads in the country and partly on stone-paved roads in the city. Early industrialization in the West brought in some vehicles in the 1910s and roads paved with slabs and pavers, followed by concrete or asphalt. By the 1920s, major cities were linked by railroads running between Chengdu and Chongqing and between Wuhan and Kunming. In 1937 infrastructures were improved due to movement of the central government from Nanking to Chongqing, which was a more stable wartime capital because of Japanese invasion along the coasts. From 1949 to 1978, as in other parts of China, Chengdu was inundated by the Cultural Revolution. With the implementation of open-door economic reform in the 1980s, the proliferation of vehicular roads began to form a new "turtle city," with layers of ring roads encircling the old inner defensive walls. With the westward drift of development in the 2000s, the ring roads and highways have been expanded.

CHENGDU: THE CITY OF HEAVEN AND WIDE AND NARROW LANES

Social Needs and Aspirations: Heritage, Urbanization, Revitalization, Lifestyle, and Accessibility

Well endowed with history and heritage, Chengdu is in a topographic enclave. The mainstream Han culture is mixed with the ancient Basu and other subcultures. This Su (蜀) State of the Three Kingdoms (三國), a large basin circled by mountains, has been a major agricultural producer of rice, pepper, tea, and other crops for the West and other parts of China. The minorities from the Northwest and Southwest would trade at Chengdu, with some settling there over time. Due to difficulties in access, the area and adjacent regions were often used as a haven during wartime.

Chengdu was a walled fortress city as early as AD 310. Chief architect Zhang Yi (張儀) decided its borders. Demolition continued with the Royal Palace in the 1960s. In the 2000s the Sanxingdui (三聖堆) and Jinsha (金沙) archaeological sites were found, indicating habitation some four thousand years ago. The name Chengdu means "becoming a city," and names like Du Fu (杜甫), Marco Polo, and Zhang Xiangzhong (張獻忠) are related. Topics like the railroad protection movement (1911), Operation Matterhorn, and Chiang Kai-shek (蔣介石) are well-known in this region, as is the bombing of Japan using U.S. B-29s. Aside from numerous monasteries, a wide range of historic towns and neighborhoods such as Jinli (綿里), Honglongxi (康隆寺), Chunxi Road (春熙路), Anren Old Town (安仁古镇), and Luodai (羅大) spread through the area, amid vernacular townscapes and architecture. A few European-style churches and buildings punctuate the area. Until the 1990s, the Wide and Narrow Lanes district was mainly for pedestrians and bicycles, with only a few cars passing by. Elsewhere, in smaller historic townships and urban edges, mahjong became popular. As part of central economic planning in the early 2000s, Chengdu decided to prepare comprehensive plans for the urban center. This included property development, revitalization of neighborhoods, and development of transportation. This was a follow-up to similar movements in the United Kingdom, where government units were reorganized to facilitate such development. In China the effort was to transform many public cultural affairs into private cultural and creative industries with a commercial perspective.

Economic Development: Western Frontiers, Agriculture, and Sustainable and Emerging Investment Showcase

In the 2010s Chengdu has a high-speed, one-hour rail link to Chongqing, with extensive rail and bus lines providing internal and external connections. The 1,247-kilometer Lanzhou–Chengdu oil pipeline from Xinjiang in the West passes through Chengdu. Coupled with this is the idea of sustainability, the development of low-carbon cities with less pollution, and higher mobility for employment. The town center is a node to be redeveloped into a pedestrian district as part of wider planning. On the industrial and technological fronts, Foxconn, with more than 1 million employees, has become the major manufacturer of Apple computers. Meanwhile, Dell has also invested substantially in building plants, as abundant technical staff is available from local universities. China has reframed its innovation policy toward an indigenous approach with these firms, where previously the government, universities, and public labs were key players. Hopefully this will generate more opportunities for the private sector. In 2012 Chengdu and the six neighboring provinces in the West had a higher GDP growth rate (9 percent) than the affluent coastal provinces and cities. With quantum leaps in virtual finance and wealth management, analysts have graded Chengdu and Chongqing highly, including for their educational and knowledge-based industries. Land- and industry-based initiatives are the Wenjiang New Town Plan (温江新市规划), covering sixteen square kilometers with some eight functional districts in leisure, health care, agriculture, science and technology, cultural industries, and the Chengdu International, Science, Education, and Arts Town (CISEAT). However, international development capabilities, as in human resources, still have to be improved. In any case, the city hopes to leapfrog to modernization and industrialization more quickly than the coastal cities.

Technology and Industry: Smart and Resilient Indigenous Craft and High Technology

Historically, Chengdu was accustomed to indigenous technologies, such as the fire heating of cold water to build the two-thousand-year-old Dujianyin (都江堰). It also developed a wide range of techniques in cultivating tea, pepper, rice, and bamboo, as well as the creation of folk crafts. The industry chain also extends to food, and Chengdu has been designated by the United Nations as a culinary city. But with recent contamination of key dairy and other food products, this area needs strengthening. With high expectations and formalized markets in the 2010s, governance, management, and ethics have become more important. During the 1940s, the U.S. Flying Tigers had some facilities in the region, using caverns along highways to store weapons and supplies. These extended all the way to Guizhou, Yunnan, and Burma. From 1949 to 1978, the agricultural sector was based in communes, which were largely labor-intensive. Productivity was unreliable and of low value. It was inadequate to support the other parts of China, so wheat had to be imported from North America. With open-door economic reform after 1978, farmers were allowed to plant their own land and thus had more incentives to increase production. Many indigenous techniques, such as gravity irrigation flow through channels and the building of retaining walls by wrapping pebbles with bamboo skins, have continued.With a population of 14 million and dozens of universities as training grounds, Chengdu is attractive to foreign direct investment, including manufacturing, high-tech, and biotech, with some industries related to its agricultural base. Aside from locals, graduates returning from overseas universities are also sought after. With natural disasters such earthquakes, Chengdu has learned to better design with nature; high-rises have been built to high technical standards.

Amid frequent flooding, rain, earthquakes, and other disasters, urban resilience and the idea of a "smart city" have become major topics. These include relatively low standards of guobiao (國標)—for example, drainage and power systems—as well as management. In September 2013, Typhoon Usagi resulted in zero casualties in Hong Kong versus twenty-five in Shenzhen. Floods have caused substantial social and economic losses in many areas. Due to proper education, activities in Hong Kong are almost automatically resumed within two hours after the lowering of signals. A key difference is the development of an effective preventive system rather than a protective and after-the-fact one. The system should forewarn people, particularly with information and communication technology.

Policy and Initiatives: National and Regional Thrusts, Clustering Synergistic Sectors, Local Pedestrianization, and Cultural Industries

In the 12-5 development plans (2010s), western China has been highlighted as a new region for development and investment. More development is planned for eastern Chengdu in Longquanyi (龙源驿), a former historic depot. Industries may include automobile manufacturing with production capability of three hundred thousand, plus production of mechanical equipment, food, and basic materials such as metal, glass, and synthetics. The following are specific development plans focused on agriculture:

- National Resource Integrative Development Three High-Level Agricultural Demonstration District (国家资源综合开发型三高农业示范片型区)

- National Agricultural Results Demonstration and Testing District (国家农业成果示范试点区)

- National Agricultural Standardized Production and Demonstration District (国家农业标准化生产示范试点区示范区)

- National Pollution-Free Fruit Demonstration District (全国无公害水果示范区)

- National Pollution-Free Agricultural Products Processing Base (全国无公害农产品加工基地)Tourist attractions are widespread, with some historic vernacular towns, transit stations with bicycle paths, and institutional campuses. A recent proposal for CISEAT for a series of liberal arts colleges included such a linkage system for several small campuses. Over and beyond would be extended paths, sports, and water-based activities along the Dujianyin, which could be used for rowing, similar to that common at U.S. Ivy League colleges.

Case Study: Wide and Narrow Lanes

Urban Design and Planning: Historical and Status Quo Model (before 1990s)

The Wide and Narrow Lanes district has a grid-matrix urban form. It goes back to the feudal monarchy and agricultural society, when the city walls were built during the Ming Dynasty (AD 1100s). The district was destroyed during wars and rebuilt many times, with quite a few barracks constructed, particularly in the face of invasion by the Mongols. During the Three Kingdoms, it was used by the Su Kingdom as a capital and was associated with historic figures like Liu Bei (刘备), Zhu Geliang (朱葛亮), Guan Yu（关羽）, Zhang Fei (张飞), and Zhao Zhilong（赵志龙）. Their opponents and partners were Cao Cao (曹操) of Wei (魏) and Sun Quan (孙权) of the Wu states, with changing roles over time.

Figure 3.3 Integrating old and new

The Wide and Narrow Lanes district was built during the Qing Dynasty (AD 1616) to house the Eight Flag reserve regiments (八旗后备营) of about one thousand soldiers. It was designed on a three-level hierarchical urban system: Dacheng (Big City大城), Huangcheng (Imperial City 皇城), and Xiaocheng (Small City小城), of which the last was to house the regiments. Some of these areas were demolished in the Xinhai Revolution (辛亥革命) in 1911 and throughout the 1950s. Together with the shooting grounds near the Bei Jiaochang (北较场) north gate and Wuding Bridge (五定桥), they were part of military systems built to defend the West. The structures, built of brick, bamboo frames, and lime mortar, have been transformed from military to mixed residential and small-business use (大雜院). The area is used for cooking, storage, and the transportation of honeycombed coal blocks. The buildings are mainly black and gray with patches of red, with the last used on signs and other cultural motifs. After the 1910s, with the introduction of vehicles and trains, the Wide and Narrow Lanes became more deteriorated and subject to considerable redevelopment pressures. However, the houses, small shops such as barbershops, and cooking activities continued after 1949. The residents were mostly older and lower-income families. Occasionally, a few salvagers lingered around to pick up used household items such as TV sets.

Figures 3.4 & 3.5 Cultural motif and articulation of architectural and landscape spaces

Modern Model: Proactive and Innovative (1990s–Present)

As part of 12-5 economic planning, cultural and creative industries are encouraged for urban development, particularly for revitalization. Within the context of "the City of Heaven" and "the Garden City," in 2003 Chengdu decided to revitalize three historic areas. The City Investment Company (城市投資公司) and the Chengdu Cultural and Tourism Group (成都文旅集團) were key investors, emphasizing an "authentic" (原實) balance of public and private purposes to showcase urban and cultural revitalization. Architects and planners from Tsinghua University were hired as consultants. The approach is to transform cultural affairs in the public realm to private civilian enterprises as much as possible. This minimizes public costs and makes good use of private capital and talent. While the profiles of enterprises may vary, investments are from both public and private sectors. Cultural entrepreneurs are invited to develop a wide array of activities such as art shows, singing, dancing, and bazaars to be held in public open spaces. Only a low fee and simple registration are needed for small enterprises and individuals to take part in the development. Social and nonprofit groups that perform dance, tai chi, and so on can participate for free.

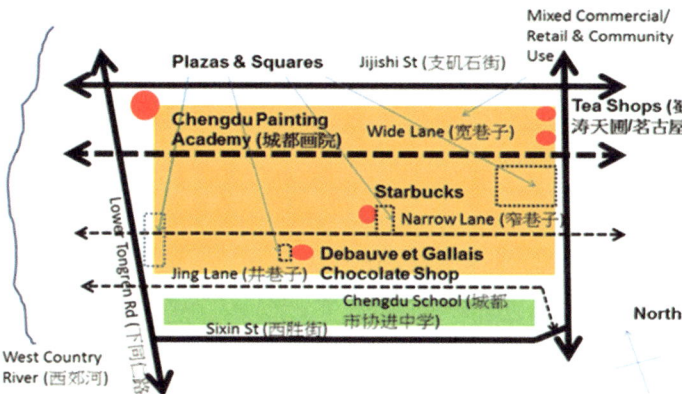

Figure 3.6 Chengdu: Wide and Narrow Lanes urban design and planning concept

Access via taxis and other cars is possible, though underground parking fees are required. The use of public transportation is encouraged. The designated cultural revitalization area is about thirty-two hectares in size. Some shops are open in front, while others are well designed and detailed to offer a come-on with hidden surprises. The architecture is mostly traditional courtyard types based on northern vernacular styles, built with gray-green bricks of high quality. With considerable deterioration such as leaking roofs, cracking walls, and inadequate washrooms, these structures have been repatched and improved with sensitive add-ons. In brief, the development model is one of adaptive reuse—that is, "old vase, new wine"—transforming the area into a place with diversity, vibrancy, and viability. However, design details are properly controlled to ensure a balance based on traditional Chinese architecture. The three parallel ancient streets have complex northern-style one- and two-story courtyard houses of about thirty thousand square meters. The investment was about RMB 600 million, with shops offering Sichuan cuisine and tea housed in revitalized buildings, plazas, and atriums. Mixed brands are based on formal contractual relationships—for example, Shiguanghua Upper Table (石光华上席), Tea and Old Horse Trail BBQ (茶马古道海鲜烧烤), Indian food, French chocolate, Starbucks, and so on. The key elements that form the image of the place are:

- Paths: Three main streets run through the axis. Districts are based on a fish-bone pattern, with a total of eight Guan streets (官街) and forty-two Soldier streets (兵街). Widths vary from eight to eleven meters. Utilitarian services to shops are provided via back lanes or after shops close.

- Landmarks: Major buildings and gates formed the General Yamen (將軍衙門). The French chocolate store Debauve et Gallais was formerly a missionary hospital. A typical residence built in 1935 was renovated to demonstrate Westernized lifestyles during the Republican years. Several hotels are centrally located on- and off-site, within walking distance.

- Nodes: These comprise the plazas and squares, with public bus terminus at both ends. The larger one is toward the west, where dancing by social and arts groups is popular.

- Districts: Shao City is comprised of a large plaza, previously General Yamen. Big City and Imperial City refer to the broader Chengdu area, but at a distance from public transit.

- Edges: The vicinity is bounded by several bypasses for vehicles. Together with other historical districts in the urban core, it was previously enclosed by walls. The interiors of renovated buildings are formal markets, while exterior spaces are informal markets such as bazaars.

Impacts and Issues

Overall the project is a success, with about 36 million visitors in the four years between 2008 and 2012. Despite financial crisis in the United States, it was able to draw 8 million visitors from there, followed by an annual revenue of RMB 400 million. It has organized itself through the Chengdu Qingyang District Government Cultural Management Association (青羊区市府文化管理员会) to coordinate various development tasks. Relocation could be done by several methods: relocating to the original building or site (回迁或原地安置), relocating elsewhere（異地安置）, with cash compensation（貨幣安置）, or outright purchase or swapping of development rights （產权買斷或調換）. However, almost all original residents were moved away from the area. Despite economic success, it has created a kind of neighborhood disassociation and gentrification.

A key infrastructural support to make this urban design plan possible is a new transportation system, with specific routing and bypasses to facilitate pedestrianization. This is also part of the larger plan for construction of mass transit and a cap on the number of cars operating along the third ring road.

The development project is a success, with cultural and creative industries integrated with a tendering system. The pedestrianization scheme has clustered the pillar industries as a precedent and showcase, as well as transforming a former dilapidated development. However, it is still very much a brick-and-mortar approach, as little information, communication, and technology have been used. Perhaps with such enhancement, its Basu cultural and subcultural elements could further be augmented on the Internet to showcase Chengdu as a future cultural and techno-pol.

KUNMING: URBAN AND REGIONAL SETTING

Changing Kunming's Urban Core: From Walled Old Street Outpost to Multiple-Grid Blocks

Kunming is in Yunnan Province and is situated in the far southwestern corner, accessible to Burma and Vietnam via an hour's drive on the highway. To the immediate north is a series of smaller historic towns such as Dali and Lijiang, renowned for their scenic beauty. Further north is Sichuan, which leads to northwestern China, including Tsinghai. To the east are Guangxi and Guizhou provinces, which share the mountainous nature of the region, with trade and other links with the Pan-Pearl River delta. To the south are Laos and Vietnam, which historically have maintained close links in trade, political, military, and other activities. To the west is the northern tip of Burma. Historically, Kunming was a remote outpost oriented toward Southeast Asia, with less contact with the rest of China. So Kunming is a historic city overlaid with Han rule and cultures, albeit with considerable autonomy. Chinese cultures of Confucianism, Buddhism, and Taoism permeate the city in many ways. As a historic walled city, it has been transformed in many ways to adapt to modernization and industrialization.

Figure 3.7 Kunming old city wall

The Case of Kunming: Spring City and Old Street

Social Needs and Aspirations: Heritage, Urbanization, Revitalization, Lifestyle, and Accessibility

The history and heritage of Kunming date to the AD 200s, when it was a prefecture city for caravan trade with Southeast Asia and India. Lake Dian (滇湖) and its vicinity were associated with a kingdom of the same name. In the Han Dynasty, it became part of the Yizhou Commandery (益州郡), but the king of Dian was left as the local ruler. Founded in AD 765, Kunming was named Tuodong (拓东) and later the Kingdom of Dali (大理). During the Ming and Qing dynasties, it was the seat of the superior prefecture of Yunnan. In the 1800s, there were several rebellions, which destroyed the Buddhist sites and the wealth of the city. In the late 1800s, the French started to build the Kunming–Haiphong (昆明-海防) railway for trade and the shipping of weapons to suppress various revolts. Mineral resources were also tapped. Yunnan had close but difficult links with Guizhou, Sichuan, and Hunan, often using mules and pack-carrying porters, though these conditions have been improved considerably in the past decade.With the Japanese invasion along the coasts in 1937, many migrated to the southwest and transformed Kunming into a modern city with a population of 140,000. In addition to the merger of universities in Peking, Tsinghua, and Nankai to become the National

Southwest Associated University (NSWAU or Lianda 西南聯大), Yunnan University was established in 1922. The universities gathered some of the best intellectuals of China and overseas at the time. While Lianda has remained low-key, Yunnan University has a few stately buildings and mature landscapes, as shown in the movie In the Heat of the Sun (阳光灿烂的日子) by Jiang Min (姜文). The road between the two campuses has been renamed 12.1 Street to commemorate patriotic activities of some of the faculty and students. In the 1950s, a large Minorities Institute was built to serve the multiethnic populations. In addition to some 4 million Han, the key minorities are Ti (彝族, four hundred thousand), Hui (回, 149,000), Bai (白, seventy-three thousand), and many other tribes. Some of the minorities have stayed in exclusive communities. During the Cultural Revolution, Kunming was a place for political exiles. With strong affinity to the mountains, valleys, and lakes, it has maintained its agricultural base. Kunming has been the provincial capital, while smaller, vernacular towns such as Dali, Lijiang, and Shangri-la punctuate the North. Overlaying the Han and other subcultures are Confucianism, Buddhism, Taoism, Islam, and Christianity. Some elements were imported from the central plains. Others came via Guangdong, Guangxi, Xigui, Tibet, Burma, Thailand, and Southeast Asia. Quite a few Western scholars moved from the coasts to form the NSWAU, before military figures such as Cai Er (蔡锷 Tsai Ao) rose against Yuan Shi Kai in the early Republican years. Yunnan and the Southwest were important places to balance national China during unstable times, as shown by generals Bai Chongxi (白崇禧) and Li Zhongren (李宗仁) from Guangxi, Sichuan, and other provinces at different times. Some of their troops were perceived as the best in China.In 1978 the coastal cities embarked on open-door economic reform, with Shanghai and Beijing picking up the momentum in the early 1990s. During this time, some policies were also spearheaded in the Southwest, along with the successful designation of Lijiang (丽江) as a UNESCO World Heritage Site (1997). In Kunming there have been concurrent efforts at tourism, cultural revitalization, and urban revitalization. A series of historic towns are planned to induce tourism, along with a policy to transform cultural affairs to cultural and creative industries, with important lessons from the United Kingdom (1997), the rest of Europe, Japan, Korea, and other places . In the 2000s, the 11-5 and 12-5 five-year plans (2006–2010, 2011–2015) take Yunnan forward to develop its culture and tourism industries, as well as related urban development and revitalization. A series of policies, plans, and studies were conducted on a wide range of heritage areas and mixed and multicultural vernacular townscapes and architecture, while enhancing living quality and job creation. Lake Dian and its coasts have emerged to be part of the city of Kunming proper.

Figure 3.8 Former Young Men's Christian Association

Economic Development: Western Trading Frontiers, Tourism, Agriculture, and Sustainability Showcase

Although Kunming has dominated Yunnan in the South economically, it lagged behind the coasts during and before the Qing Dynasty. In the early Republican years, it became an industrial and manufacturing base for Chongqing, which was the wartime capital. An elaborate system of underground caves, barracks, and factories was built as a last redoubt, in case Chongqing fell. The British and French also had military offices in Kunming during the Japanese invasion to liaise with their troops in Southeast Asia. The only linkage to Southeast Asia and the rest of the world was through the winding Stilwell Highway, which was built with six hundred thousand wartime laborers. It was the gateway for resources and the last resort for lifeline supplies, as the coasts were blockaded by the Japanese. It has always been externally oriented.

The U.S. Flying Tigers had their facilities and service centers here, acting as a major unit to help China fight the Japanese. The unit also flew considerable amounts of critical supplies such as weapons and medications. With Chongqing becoming the wartime capital, mining and the manufacturing of electronics, copper, cement, steel, paper, and textiles continued. After 1949 large iron and steel factories were built. In the 1960s and 1970s, economic developments almost came to a standstill due to the Cultural Revolution.

In the early 2000s, the development of the western regions (西部大开发) was introduced, covering the provinces of Gansu, Guizhou, Qinghai, Sichuan, and Yunnan. In 2002 the construction of the West–East Gas Pipeline began, followed by the returning of grazed land to grasslands. In the mid-2000s, Kasikorn Bank and Krung Thai Bank were set up by Chinese from Thailand, with Princess Maha Chakri Sirindhorn studying Chinese culture and promoting friendship. In 2005 the Mekong Subregion Summit was held, with China donating US$1 billion to spearhead a Chinese Association of Southeast Nations (ASEAN) free trade area. In 2007 the Ministry of Finance invested RMB 280 billion in the West to support key projects. Infrastructure and aviation have been greatly improved, and the World Horticultural Expo has become a regular event. In addition to balancing the affluent coasts, narrowing income gaps between urban and rural and relieving pressures on urban services like housing, schooling, and hospitals are important tasks. Kunming has become a trading hub and gateway to Southeast Asia and part of the PPRD 9+2 regional plan, focusing on agricultural commodities such as tea, horticultural exports to Europe with high margins, and minerals such as jade and precious metals. In the early 2000s, Kunming began to revitalize its urban core. While marketing is a 360-degree endeavor by everyone from major conglomerates to individual freelance practitioners, both formal and informal enterprises are encouraged. These may range from Parkson and Carrefour to KCF and Starbucks. The cultural creative industries permeate more than ten blocks, with frontage stores along Old Street dominated by branded fashion, jewelry, and luxury goods. Brands like Luk Fook and Chow Tai Fook are examples. More than one hundred retail stalls are present on the fifth floor of the New Millennium Hotel and Kunming Department Store (新纪元酒店及昆明百货大楼) at the Nanping Street–Sanshi Street junction, complete with a testing center. They coincide with Kunming being a key chain to set certain standards—for example, guobiao (Chinese national standards). The second and third clusters planned are department stores and restaurants. The towers around the plazas are clustered with hotels and major offices, with access to public buses and drop-off areas. Amid open-door economic reform after 1978, Kunming could be a late entrant. But it has moved forward with appropriate policies and incentives, particularly for direct foreign investments from Hong Kong and elsewhere. The economic pillars are agriculture, tobacco, mining, hydroelectric power, and tourism. In 1992 the Ruili (锐利) Border Trade Economic Cooperation Zone, the Wanding Economic Cooperation Zone (畹町, six square kilometers), and others were set up for tourism and airport development with Vietnam. In 1999 Sino–Burmese trade accounted for US$55 million and 77 percent of Yunnan's foreign trade. In 2007 GDP of the primary industries of Yunnan rose to fourth place nationally, after Hainan, Guangxi, and Sichuan, though Yunnan is still dominated by agriculture. Other examples of economic zones for experimental border trade and tourism are in Yuxi, Dali (大理), Chuxiong (楚雄), Songming (嵩明), Hekou (河口), Lijiang (丽江), Cang Mountain (苍山), Xishuangbanna (西相版納), Tengchong (腾昌), Yangzonghai (阳宗海), and Fuxian Lake (抚仙湖).In 2011 Zhang Yimou was invited to spearhead a branding score. He used a colorful female dance troupe, with layers of Yunnan landscapes as backdrops. These are observable on the Internet, Hong Kong mass transit, and other public places. They have become some of the vivid images for Kunming and the area, while Old Street (老街) is becoming a catalyst for development in many ways. With a total urban population of over 6 million, Old Street has economy of scale and well-networked urban and regional sectors.

Technology and Industry: Indigenous Craft, High Technology, Smart and Resilient

With a mixed Han population and dozens of minorities and with mountains, rivers, lakes, and farmland dominating the countryside, the valleys and flat areas have been developed as urban centers. Aside from the economic pillars, the villages have maintained indigenous arts and crafts (民族工艺), with emphasis on cultural content, such as handmade costumes, jewelry, silverware, gifts, and souvenirs. The production centers have ties with urban markets, as reflected in the bazaars of Old Street before the 1980s. Kunming was a temporary hinterland to the wartime capital Chongqing and a regional cluster to fight the Japanese invasion. The Flying Tigers were based here, with related service centers at Lianda, plus development of certain manufacturing industries. However, when the Kuomintang decided to withdraw to Taiwan in 1948–1949, these were not developed further. At any rate, caves along highways in the region still show their function as storage and military-based technological development areas during the wars.

After the 1990s, the scientific and technological capabilities of certain enterprises, such as Yunnan Baiyao (云南白药), were extended to become more entrepreneurial. It has actively participated in Old Street redevelopment by making major donations. In and near the city, many science and technology parks were created. These include:

Kunming Economic and Technological Development Zone (nine square kilometers): Located in east-central Yunnan Province, with a focus on tobacco processing, machinery manufacturing, electronic information, and biotechnology

Figure 3.9 Old Street pedestrianization

- The Qujing (曲靖) Economic and Technological Development Zone (106 square kilometers): Targeting a population of four hundred thousand as an economic and technological center in the 2010s. There are several economic and technology parks in the area.Urban transport is somewhat chaotic, with a mixture of vehicles of different sizes and speeds, inducing defensive driving and making road crossing difficult. Pedestrians are in danger of being run down, even if properly crossing with traffic lights at intersections. A major public bus terminus near Xiao Ximen (Little West Gate 小西门) is dominated by cell phone shops, but with few restaurants and other supporting services. Although aligned with a few major hotels, this well-located subcenter lacks diversity. While many Chinese cities are striving to become "smart," with higher levels of resiliency, the principles of information provision, tourism friendliness, directional signs, and warning systems seem to be inadequate.Yunnan has some twenty-four higher educational institutions, among them the historic Yunnan Normal University (Lianda) and Yunnan University. Others include Yunnan Academy of Agricultural Science, Kunming Medical University, and Yunnan University of Traditional Chinese Medicine. They produce some of the key talent to satisfy local development needs. As with many schools across the nation, matching and mismatching of graduates is an issue. Some prefer to go to coastal cities, Hong Kong, or overseas for more internationalized education. Minorities are still poorly educated, despite the national policy on rural–urban migration. With urban land constraints, many new campuses and facilities have been built in Chenggong (呈贡), a new town to the south of Kunming.

Policy and Initiatives: National and Regional Thrusts, Clustering Synergistic Sectors, Local Pedestrianization, and Cultural Industries

Historically, Kunming and Yunnan have had a strong sense of autonomy for many centuries, with prominent figures like Zhu De (朱德) and Cai Er (蔡锷). Some soldiers have retired in the area. Others are across the border in Burma. Recapturing the more distant past were Lin Xiexu (林则除), Wang Xi (王炽), and Tang Jiyao (唐继尧). Nevertheless, policies and initiatives in the Republican period were for military and national unity. As with infrastructure, weapons and supplies were key. Often hard labor was mobilized with primitive tools and equipment to build major projects for national needs—for example, Stilwell Highway. After 1949, Kunming and Yunnan were still externally oriented—trading with Burma, Thailand, and Vietnam. The last was instrumental when it took the front line to fight the Americans. During the Cultural Revolution, economy was at a standstill. After the 1990s, the 11-5 and 12-5 plans were instrumental in spearheading economic development of this remote outpost. In 1952 the old city walls were demolished to build the Qianlian Lu (Youth Road 青年路) in the east to facilitate vehicular traffic. Many old buildings and districts such as the One Seal House (一颗印) were also rebuilt, albeit for utilitarian uses and with basic reinforced steel. In the late 1980s, with further deterioration of many old building with heritage value in downtown Kunming, revitalization plans were prepared for more than ten blocks, or ninety-five hectares, including Zhenyi Road, Sanxi Street, and Jinbi Road (正义路, 三市街, 金碧路). The urban policy is integrated with its resource base, such as minerals, precious metal such as jade, agricultural commodities like tea, and mainstream and subcultural industrial chains (tourism and expos linked to other products). Tourism and hospitality has become a key sector, relating to many other realms, as pleasure and business trips can be mixed into one. The city and the province have worked together to maximize their efforts, with the planning bureau already expanded several times to include a brand-new exhibition hall to showcase their work. It is also clustered with the Kunming Museum and a Zengqing Taoist monastery (真庆觀), a few blocks or a ten-minute bus ride from Old Street.

Case Study: Kunming and Old Street Pedestrian District (Zhang Road, Sanxi Street, and Jinbi Road—The North–South Axis)

Cultural and Creative Industries

By focusing on the theme of a sustainable, ecological, garden, and spring city, the urban revitalization is integrated with cultural and creative industries. Investors, enterprises, and participants have become important cultural and economic resources in the development process by capturing tangible and intangible values. In 2013 the Yunnan Province Cultural Industry Development Team Office (云南省文化产业发展领导小组办公室) was part of these efforts in many ways. Cultural affairs in the public sector become cultural industries in the private sector. Meanwhile, educational programs are set up for talent development to enhance a sense of belonging. The Huhua district (五华区) was planned with some fifteen pilot cultural industries. A Yunnan Cultural Industry Expo was held in August 2013, with an area of some sixty thousand square meters. In 2011 this was an outcome along with completion of the third ring road, which together with the second diverts traffic to bypass the urban center. This maximizes the effects of pedestrianization by emphasizing rituals, lanterns, local handicrafts, and human activities in the city and among indigenous cultures near Dian Lake.

Figures 3.10 & 3.11 Activities along the pedestrian zone

Urban Design and Planning: Historical and Status Quo Model (1978–2000)

The city walls were built during the Ming Dynasty (AD 1362–1644), beginning in AD 1382. Several villages were developed for soldiers within the boundary; today they have become featured *toscens*. In the next six centuries, some forty streets were built inside the walls. With a growing economy, roads and streets extended outside the walls in the late Ming and Qing dynasties. The Old Street district began development in AD 1119 and already has some nine hundred years of heritage. In 1952 the government demolished the walls and used the bricks to build a road linking North and South.

Figure 3.12 Zhongai Gate (Righteous and Love)

This generally aligns with the first ring road, with some local names surviving as places—for example, Xiao Ximen (小西门), Beimen Jie (北门街 North Gate Street), and Qinglian (Youth Road 青年路), with the last located along the east wall. After 1978 Old Street pedestrianization is located in an area east of Zhengyi Road (正义路), west of Wuyi Road (五一路), south to Jinxing Street (景星街), and north to Renmin Zhong Road (人民中路)—a total of about ninety-five hectares and more than ten blocks. *Modern and Innovative Model (2001–Present)*Land and property development took the form of a public–private partnership based on a parcel tendering system administered by the city government. To date, major developers are Vanke (萬科), New World, Jijiang (之江), and others. They also package their retailers. There is also a supplementary licensing system for freelance small and individual businesses. Considerable historic areas and businesses have been preserved along the edges of the urban core, with some reaccommodated in historic buildings. Others, with lower and more affordable rents and prices, continue to operate within walking distance. The city maintains a geographic information system (GIS) for planning and environmental monitoring.

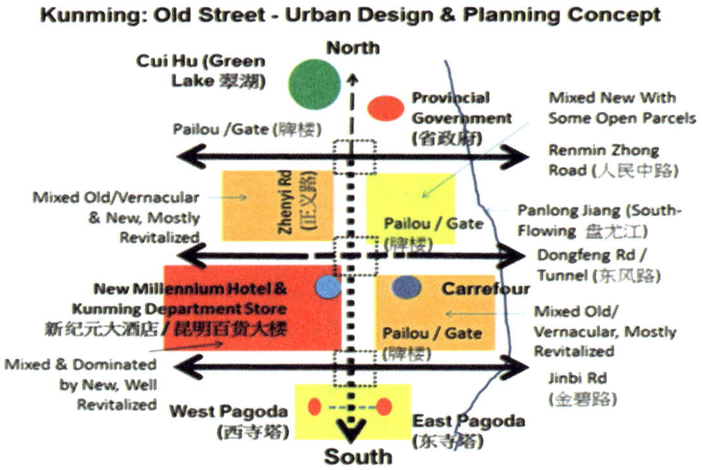

Figure 3.13 Kunming: Old Street urban design and planning concept

A key infrastructural support to make this urban design plan possible is the new subway system, which began its first phase in 2010. By 2013 only Line 1, from the new Changsui Airport (長水) in the Northeast to the East Bus Terminus for a transfer to the pedestrianized area, had been completed. A total of some nine lines are forthcoming. With a budget of some RMB 300 billion, which is ten years of the city's total financial income, the city is having problems meeting project payments. To better coordinate the project and technical and human complexities, thirty-four-year-old Li Liang (李亮), with a doctorate degree in planning from Tsinghua, replaced the vice mayor in 2011. In terms of urban design and planning, several consultants have been employed. Concurrent research has been done by the Sustainable City Programs of the United Nations. It was also in on the urban design and planning of some forty-two hectares for the core urban area elsewhere in the early 2000s. The following are elements of the image of Kunming:

- Paths: The main street runs through the axis, linking Cuihu Lake (north 翠湖) to the two pagodas (south). The districts were formed on the east and west sides of the axis.

- Landmarks: Major modern architecture includes the New Millennium Hotel and Kunming Department Store (新纪元酒店及昆明百货大楼). Near Sanshi Street are major administrative and historic buildings, including provincial government buildings, a temple, and a pagoda. Occasional smaller historic landmarks include the former residence of Nie Er (聂耳) and his statue. He was composer of the national anthem for China (义勇军进行曲).

- Nodes: Major activity areas are the plaza facing the New Millennium Hotel, where public transportation comes in on the east and west ends, while squares and gates are located at Jinbi Square (金碧广场) and Jinma Lane (金马碧鸡坊). These last two were made possible with donations from Yunnan While Medicine (云南白药) and others.

- North: The northern terminus of the main road coincides with the old city wall gates, starting from Cuihu Park (翠湖公园), which has an area of twenty-one hectares, with lotus, ducks, hotels, teahouses, a zoo, and cultural lanes (文化巷), which are popular with youths.

- South: Three major squares are located along the north–south axis leading to the Dongji (东寺East Pagoda) and Xiji (西寺 West Pagoda), with Nanping Road (南屏) as an intermediate linkage along Dongfeng (东风). They can be dated back to the Tang Dynasty (AD 824–829) and are often used for tai chi and exercises.

- West: A large antiques and cultural industry market, with the bazaar-type Jingxing Birds-Flowers Market (金城雀乌花卉市场).

- East: A series of small businesses are housed in revitalized blocks of historic buildings of two or three stories—mostly restaurants, arts and craft shops, and so on. They include Yuanhexiang (云和祥), Aiyechun (蔼若春), Liaodong Congee (老东粥皇茶餐厅), and Huruen Fashion/YMCA (胡润服饰/基督教青年会, 1933). The last was designed by architect Li Jinpei (李锦佩), a graduate of the University of Pennsylvania, who also designed the Nanking Sun Yatsen Memorial (南京山陵).

- Edges: These were previously walls and now have become layers of edges around the new pedestrianized blocks. Inside the modern buildings are formal markets, while outside are small informal markets, mainly linear bazaars. Several clusters are apparent—for example, general shopping, food and culinary, banking, retail, and jewelry such as Jinxing (景星珠宝城). The old houses are preserved and refurbished, and adaptive reuse is realized as much as possible, thus providing contrast of old and new. Some of the highlights of modern retailing are international brands CK Jeans, Teenie Weenie, and Best Seller from Denmark. Those integrating Chinese traditions and the vernacular are Lazy Man Coffee (懒人咖啡) and various teahouses and stores along Jinshing Road.

Impacts and Issues

Performance

General urban design and planning principles involved mixed use and clustering. Public transportation is external but could go right up to intersections near the central plazas, while limited access is slow for buses and drop-off is low speed. In 2005 average daily pedestrians numbered thirty thousand, with one hundred thousand on holidays. On a three-day holiday in May, the total count was 1.2 million, or four hundred thousand daily.

Relocation

The plan involved the relocation of some nine hundred units, and about one hundred are remaining. There are two approaches to redevelopment: one is to "demolish the old, build the new"; the other is to "build and revitalize simultaneously." With plans staged for longer than ten years, compensations have been rising. Among many developers and stakeholders, Jijiang Real Estate (之江置业) has been active with its Kunming Historic Street District Protection and Development Project (昆明街历史街区保护建设项目). The district would involve some eighteen streets, with historic brands such as Qiangwang Piao Hou, Tong Xinfeng, Xiao Yenkwei Xiong (former bank units 钱王票号、同庆丰、小银柜巷), Ma Family Estate (马家大宅), Fulin Tang, Fuhua Yuan, and Jian Xinyuan (福林堂、福华国、建新园). A Kunming Old Street Study Group (昆明老街研究会) was also formed to facilitate dialogue, research, and development of various tasks.

Planning and Development Leadership

In 2011 Kunming appointed Li Liang (李亮) as chief of the planning bureau. He replaced the vice mayor and director of the Yunnan Provincial Government Finance Office (刘光溪) to handle all complex tasks and projects, which require strong knowledge-based training and skills. With experience from major coastal cities, it is anticipated that some of the urban design, planning, and developments can be improved over time.

Finances and Project Delays

A loan of RMB 300 billion is to finance mass transit. But this equals ten years of the city's financial income. In 2009 this loan was made through the China Development Bank, with the city investing RMB 63.5 billion in rail transportation. However, "The gap between financial income and outlay has widened, a problem aggravated by the stable monetary policy of the central government, which increases the difficulty of financing companies in obtaining funds and boosts loan costs" (www.BambooInnovator.com). The plans comprise some nine mass transit routes and five railway lines totaling 562 kilometers in length. At this time, only Line 1 of the mass transit system is substantially completed, while the last short section from the new Changsui Airport leading into the urban center is still missing; passengers have to transfer to a bus or taxi. The transit network is the largest infrastructural project in Yunnan. Thus cultural creative industries are integrating with revitalization among the broad realms of property and land development. While striving for downtown pedestrianization, Kunming has a vision for an emerging middle class, progressing toward affluence by catching up with coastal cities. With an exceptionally favorable climate, the city and environment offer enjoyment, safety, and comfort—"more for people, less for cars." The Old Street revitalization has clustered economic pillars as a catalyst, thus preserving historic "green brick" architecture and enhancing social well-being. Despite some issues on project financing and human resources, it is capturing added value on cultural heritage and is in line with policy. While people have high expectations for modern comfort with empathy for the historical past, the preservation of old brick and mortar and the contrast with modern steel and glass would be a positive approach. Over and beyond these physical aspects, it remains to be seen how smart and resilient the city can be, as many other cities have already advanced virtual, information, and communication technology upgrading as a competitive measure.

Case Summary

As part of the PPRD 9+2 thrusts, Kunming and Yunnan have achieved considerably in development efforts. Development plans are active in particular with Southeast Asian nations across the border. However, some key issues on the Old Street urban core, such as leadership and finance, have remained. The latest policies and updates to strengthen Southeast Asian trade will make Kunming and Yunnan more economically and financially independent. However, leaders with appropriate international and regional knowledge and skills are difficult to find. Whether income could be strengthened to cover infrastructural loans remains to be seen. The training of local talent does not seem to be catching up with development needs, as leaders have to be transferred from larger and better-developed coastal cities, or perhaps even from overseas (haigui or returning turtles海龟). With quick and intensive Western development thrusts, minorities have remained less educated, implying that changes in residency and mobility policies may have to be considered.

CONCLUDING OBSERVATIONS

The Chengdu Wide and Narrow Lanes is a compact block based on a grid-matrix urban form. It is based on a three-lane orientation amid the old urban center of a series of former Qing barracks in a historic city, all advancing with better talents and leadership with economies of scale. Old Street of Kunming, an expanded district from the central core, is based on the north–south axis of the historic walled city. Even with gaps in leadership and project coordination, it is capturing much cultural heritage as intangible benefits. For both cities, the issue of leaders with international exposure has remained, while they bear imprints on Western development fronts. The implications for the neighboring six provinces are apparent, if they wish to learn these lessons. Meanwhile, they apparently have contributed to nation building for China as it faces the challenge of maintaining economic growth at 8 percent. Over and beyond is to maintain political and social stability and to improve governance and legal systems to better manage corruption and ethical issues.

BIBLIOGRAPHY

Fingerhuth, Carl, and Ernst Joos, eds. *The Kunming Project: Urban Development in China—A Dialogue.* Berlin: Birkhauser, 2002.

Lynch, Kevin. *The Image of the City.* Cambridge: MIT Press, 1960.

施惟达 主编 (2006)《昆明文化产业发展纪实》, 云南大学出版社。

申静书、沈一 (2011). 「城市历史街区空间空间形态分析 - 以成都市竞窄巷子与例」, 四川大学建筑及环境学院,《安徽農業科學》, Journal of Anhui Agricultural Science, 39(4):2360-2361.

干希贤 (2000). 「昆咀历史文化各城的及脉景观特征」,《云南社会科学》第4期。

李森、宋钰红、卢显伟 (2008). 「商业步行街的比较研究 - 以昆明市南昆屏街与东寺街为例」《中国西部科技》, 5月(下旬), 第07卷,第15期总第140期。

百度文库、王志纲 (2011)《宽窄巷子最成都 - 宽窄巷子的传承与创新专题研究》。

百渡文库 (2013).《一个城市的历史、现状和未来 - 昆明的故事》百渡文库、黄靖、古红楼、清华大学 (2005)「城都窄巷子文化保护叵保护与更新设计」,《设计论祝说 - 作品评圻》, Project Review。

Chapter 4

Artistic Creativity in Modern Art Education: The Case of the Shanghai Art College, 1913–1937

Jane Zheng

Artistic creativity has always been a controversial issue in modern Chinese art education. On the one hand, the emergence of art schools in China in the late nineteenth century is viewed as an innovative endeavor. In particular, Liu Haisu has always been viewed as "the champion of the new art movement in Shanghai."[2] On the other hand, scholars are frustrated by the discovery of few creative elements in the artworks of teachers and students in modern Chinese art schools.[3] In reality, the contradiction between school education and student creativity is rarely perfectly settled, and schools have been a critical focus for suppressing the creativity of students in the West from the eighteenth century up to the contemporary era.[4] This question becomes even more complicated when the school of Liu Haisu is involved; Liu liked to use "creativity" to characterize his school.[5] In this article, I argue that two categories of creativity existed in Shanghai Art College. At the school leadership level, creativity was a response to the "new culture thoughts" (*xin wenhua sixiang*) advocated by the Education Ministry in the early 1920s. At the level of school instruction, the literati's way of artistic creativity continued, but it was applied to a cross-cultural and interdisciplinary educational context. In both senses, Shanghai Art College fostered new artistic creativity in art education.

STIMULATING STUDENTS' INITIATIVE IN ART LEARNING

Archives show that Liu Haisu's advocacy of individual creativity in Shanghai Art College was a response to the aesthetic education thought of Cai Yuanpei in the New Culture Movement, together with the school's advocacy of the artistic personality of students (characterized by "creativity"). In a historical context, "the strength of the individual citizen and the cultivation of a sound personality have been recognized to be a new focus of the education policy and media discourse."[6] Reformers attacked "the rigidity of earlier teaching structures which had overlooked the development of individuals."[7] The goal of the reformed national education system was to foster "individuality."[8] To follow the official advocacy, Liu Haisu reduced the rigidity of course selection and the intervention of teachers in the interest of students in a timely manner by revising the policies of the school. Liu Haisu liked to show off this endeavor. In practice, this policy stimulated the initiative of students to study art.

First, Shanghai Art College had a free study atmosphere, as Michael Sullivan noted.[9] However, the free study atmosphere refers to loose class discipline. Wang Geyi recalled that many students registered in his class, but not all attended. Moreover, several students paid tuition but showed up only occasionally during the year. Sometimes students in other classes or departments would sit in on his courses. He concluded, "Anyway, I taught my courses and students picked up what they like."[10] To allow more freedom, in 1931 Liu Haisu considered curtailing the length of the school schedule by a third every day.[11] Furthermore, Shanghai Art College presumably had the loosest in-class discipline among all officially registered art schools (both national and private) in the 1930s. The Ministry of Education observed, "The inspector reported that many students came late to class. He was told that class begins at 9:00 a.m., but some students attended class at 10:00 a.m. This shows that the school has a rather loose management system and it should be tightened up."[12]

Second, the school emphasized the importance of the spontaneous creativity of students in its educational syllabus. The school syllabus in the 1920s stated its purpose as follows:

The pedagogy for Year One is based on developing personal characters. Teachers give detailed guidance and explanations, only instructing the rudimentary principles and knowledge; they do not enforce students to follow certain fixed painting directions. Teachers of the course about perspective, projection and various painting methods attempted to stimulate students' passions. ... In color studies (*se cai xue*), the tone is decided by students' individual feelings and interests. It is not limited by teachers' methods.

In Year Two, emphasis is placed on spontaneous expression as well as a comprehensive study on "form," "color," and the expression of emotions.

In Year Three, emphasis is still placed on expression at painting practice courses.

In Year Four, the emphasis is self-expression.[13]

This syllabus shows that the initiatives and personal expression of students were part of the educational program.Third, the school permitted much personal freedom for expression in art studies. In 1941 Thomas Munro wrote, "Art teachers who use the term had the idea that students' artworks are 'creative' if they are spontaneously conceived and executed by the student, not done in accordance with the teacher's guidance."[14] The most remarkable policy was to allow Western Painting Department students to select the teacher they wanted to study with. Wang Geyi recalled that he taught a flower course but that not all the students followed his teaching method. Both Cheng Shifa and his wife, Zhang Jingqi, were students in his class at that time. Zhang was clever and diligent, and she seriously followed all the course instructions and did all the assigned homework. By contrast, Cheng liked to paint as he wanted. Wang stated, "I never interfere with students' painting interests. On the contrary, I particularly cherish their creativity."[15] Efforts were made to permit much freedom in examinations. In 1924 school staff decided at a meeting that if paintings were not finished in class, students could take them home to work on, but that they should submit at least one piece of work every week for the comments of the teacher in class.[16] Students were allowed to submit their work after class for examinations before or on the due date. Thus only the requirements for size and subject matter were set for the final graduation work of the students. Students had the flexibility of choosing the theme and artistic form.[17] In the 1920s Liu Haisu even tried to cancel examination grading because he believed that scores hindered personal artistic expression.[18] As shown above, class discipline was loose, individual creativity was integrated into the educational program, and personal preferences were respected. With much freedom in art learning, "individuality" was advocated as a modern concept by the school headmaster.

CREATIVITY OF MODERN LITERATI IN CROSS-CULTURAL AND INTERDISCIPLINARY BACKGROUND

The "creative education" policy of Liu Haisu in the school is similar to any other slogans or issues he created (for example, the nude model incident). The policy was in line with the advocacy of the government—that is, to cast "modern" color onto the school. However, creativity in the sense of allowing students' self-exploration often contradicts academic education. If students do not follow the directions of teachers, why do they go to school? The policy indicates that if students follow directions, they are not creative enough. Therefore, limitations always exist in school education, which is also true for Shanghai Art College. As Cai Ruohong experienced, teachers might not intervene in what students painted when they did not follow directions, but they did not give constructive advice or encouragement either.[19] Teachers had their own understanding of artistic creativity, which I argue is the preservation of the literati's way of artistic creativity. However, literati creativity was modernized in school education because this methodology was applied to art genres other than literati painting. The "literati's way of artistic creativity" is consistent with the main features of high culture. Mathew Arnold understands high culture as a science that pursues rigor and comprises a set of rules rather than subjective expressions, and the height of high culture can be attained only by reading, observing, thinking, and self-reflection.[20] Following this logic, masters in the literati tradition view creativity as something embedded in a solid understanding of artistic laws through learning, based on producing ideas or artistic expressions to strictly meet the laws.[21] Thus creativity in literati painting[22] is not separate from the fundamental process of learning; in other words, creative ideas do not spontaneously pop up and can hardly be acquired by luck. The paintings of Song and Yuan masters are the art pool for literati painters to absorb artistic ideas and expressions. Dong Qichang emphasized that every stroke should show its source and tradition (for example, which master's stroke style was followed).[23] In addition, the breadth and depth of learning as a scholar-artist is emphasized. Literati painting is interconnected with other art forms, such as calligraphy and seal-cutting and has a wide cultural background that involves the disciplines of literature, history, philosophy, and poetry, among others. The unique characteristic of creativity in literati painting, which is different from other forms of high culture, is the overwhelming significance attached to the conceptualization of ideas for expression rather than the artistic language or techniques for representation. Susan Bush pointed out that "representational aspects are devalued."[24] She also made a cross-cultural comparison, stating that "self-expression in the West is often seen in romantic terms as the solitary struggle of the artist with his material but Chinese scholars' painting was an artistic form of expression in which the personality of the maker was revealed."[25] Therefore, any artistic representation (for example, lines, color, and composition) in literati painting is not valued. Lu Yanshao claimed that only 30 percent of literati painting is about drawing;[26] 70 percent lies in the ideas and moral perfection of the artist, and only 30 percent is about artistic representation. In a strict literati sense, innovation in representation can be far lower than 30 percent in importance. Overall, the standards for assessing the quality of literati painting highly value the creativity of conceptions in the "mental state"[27] instead of artistic representation. The instructions of masters are typically to "let the conception precede the brush work; whatever is ample inside takes shape outside" ("yi zai bi xian; zu yu nei, xing yu wai").[28] Creativity in the aforementioned sense is stressed as a critical issue in judging the quality of a painting. Dong Qichang observed, "Those who study the old masters and do not introduce changes are garbage that should be discarded at a fence. If one follows the models too closely, one is often far away from the principles." This observation explains the standards and importance of the artistic creativity of literati painting. The following elaborates on how the same set of rules of creativity was applied to the modern art education of multiple art forms at Shanghai Art College.

TRAINING METHODS: "READ TEN THOUSAND BOOKS AND TRAVEL TEN THOUSAND MILES"

As creativity in literati painting entails mastery of artistic laws through broad and deep learning, the high quality of creativity relies on reading, learning, and thinking broadly and deeply. Accordingly, guiding and assisting students to learn hard from masters in ancient dynasties turned out to be the major training method for fostering the creativity of students at Shanghai Art College regardless of the art form and genre in which students specialized. Students were urged to study hard to have solid learning. In the Western Painting Department, students were expected to be diligent in raising questions, and teachers were to be diligent in giving comments.[29] The biographies of students and their comments on one another's works also show that diligent students received compliments from teachers and were considered superior by other students. Li Xiuqiong, a female student from South China, studied both music and painting. She worked hard on the piano every day and decided to pursue her study in Shanghai. Her teacher highly praised this learning attitude, stating, "It is really commendable that she is never tired of studying."[30] Yang Xinqi, another student in the Western Painting Department, was said to be full of "brute force" (*man qi*):

He is a nice person; his brute force is shown in his studies in that he never slacks off. He likes wood cutting and exerts his strength most at this craft. He likes cartoons and strains every nerve. This is also true with black and white sketch (*su miao*), oil painting and others. In the mid-spring, when other students sleep soundly, only he keeps working hard unflaggingly without blinking. Once he sets his goal on a research, he does not care about anything else.[31] A comment from a student in the Chinese Painting Department was as follows:

Shen Zian ... lost his parents in childhood, becoming studious and independent. He began to study painting in his childhood. ... In 1932, he came here to study at the Shanghai Art College. He achieved a high artistic level. His works show a pure antique and free mood, high above the vulgar without any stale taste. The composition is precise and complete, which has long been praised. He is also good at various musical instruments. Now he is going to graduate. We are looking forward to seeing an accomplished artist soon.[32] Chen Yaozheng, a deaf-mute student, was respected by his classmates because he had strong faith. Chen said, "Despite being deaf-mute, I should equally work hard for all I am worth."[33] In the comments and biographies composed by students for each other, "work hard" and "continue to work hard" are the most popular compliments and encouraging words.[34]

BREADTH

Admittedly, traditional Western art academies in France and Japan also required students to work hard to achieve mastery of painting techniques. The uniqueness of training for creativity at Shanghai Art College lies in the emphasis on a broad cultural vision. However, the school expanded both the scholarship foundation (the variety and number of masters for students to follow) and cultural background.First, the school tried to assist students in developing a broad understanding of both Chinese and Western painting traditions. Both the Chinese painting program and the Western painting program urged students to take courses in the other program to enrich their knowledge of the complementary cultural background. Students who majored in Chinese painting learned not just Chinese painting; students who majored in Western painting studied not just Western painting. In 1928 elective courses in the Chinese Painting Department[35] included Western painting theory, Western painting practice, appliance art, design, art history, philosophy, and foreign literature. Elective courses in the Western painting program included courses related to Chinese painting, such as Chinese painting theory, Chinese painting practice, Chinese painting history, and Chinese literature.[36] In 1937 speedy sketch for figures (*su xie*) and composition were listed as elective courses in the Chinese painting program.[37] In the Western painting program, elective courses included poetry, calligraphy, and archaeology.[38] Culture-related courses were also given to students who majored in applied art. In the design and architectural decoration programs, half of the courses were general cultural courses, such as foreign languages, Chinese literature, aesthetics, art history, Chinese art practice, and Western art practice, in addition to appliance-assisted painting (*yong qi hua*), industrial art, and pattern design practice.[39] Liu Haisu stressed the importance of studying art movements in different countries to prevent isolation from the rest of the world.[40] Lu Yifei drew on the concept of the "three general understandings" (*san tong*): vertical understanding (*zong tong*), horizontal understanding (*heng tong*), and inward understanding (*nei tong*). *Zong tong* refers to tracing back to the ancient era to explore the thoughts of masters; *heng tong* suggests looking at different cultures and various sectors in art; and *nei tong* requires students to assimilate what they have learned through self-reflection, traveling, interviewing, and reading.[41]

Second, Shanghai Art College also faithfully practiced the instruction of Dong Qichang about learning from nature in addition to cultural studies.[42] Neither Dong nor professional scholar-artists in the Republican period followed this instruction.[43] Wu Hufan was an example. He was a landscape painter, but he rarely stepped out of his house in Shanghai to investigate or paint real nature.[44] By contrast, at the Chinese Painting Department of Shanghai Art College, students joined the Outdoor Life Drawing Team (*lü xing xie sheng dui*) in 1925. This issue was discussed at a school meeting, and faculty decided that students in the Chinese painting program should join the team in a trip to West Lake in Hangzhou.[45] In 1926 Chinese painting program and Western painting program students embarked on a three-week trip.[46] In 1935 the importance of the outdoor life drawing tour for students of the Chinese painting program was emphasized again. The director of the department prepared well for the trip.[47] What deserves attention is that life drawing in the Chinese Painting Department was not a reform because of Western influence. Yu Jianhua stated that Chinese paintings in the ancient period were all life drawings (*xie sheng*), as recorded by Xu Xi and Huang Quan (903–965). These model paintings became an alternative for painting students only later and led to repetitive copying.[48] Life drawing in Western painting refers to drawing real objects: a mountain or water, a tree or a stone, a flower or a leaf. The process of drawing real objects involves view selection, designing composition, tailoring, and recognition, but it culminates in representation. Repetitive copying, however, is the opposite. For example, to paint a landscape, the painter should travel through the landscape first, obtain an impression, and paint what is the most impressive and outstanding, absorbing the essence and discarding the dross. At first glance, the painting might be a picture of a certain mountain, but the painted image is not exactly the same as the real mountain under careful examination: it shows the spirit and characteristics instead of detailed realistic landscape. This is also true of flower and bird paintings. Observe their postures, keep them in the mind, and then you will be able to complete a painting without stagnancy.[49] In addition, the school also provided students with a broad range of activities (for example, school art exhibitions, art competitions,[50] and art society activities). Outdoor exhibitions were held at tourist destinations at the end of outdoor trips in Hangzhou in 1923 and 1933.[51] Almost every year, three or four art exhibitions, such as the school's achievement exhibitions (*xue xiao cheng ji zhan*) and research institution painting exhibitions, were held. In 1937 the Twenty-Second Graduate's Association at Shanghai Art College held its pattern design exhibition. Around one hundred artworks were displayed. In addition to works by students, design references collected from Egypt, South Asia, Europe, and America were also exhibited. The exhibition was extended for three days because it became popular among the public.[52] Compared with the private studios of masters, the art school exposed students to a broad source of culture, nature, and society. Shanghai Art College retained the literati's way of fostering aesthetic creativity and applied it to a broad range of subjects across the following cultural backgrounds: working hard to achieve depth, studying broadly for interdisciplinary and cross-cultural understanding, and enacting personal cultivation and reflection through the life drawing team. The modernized artistic creativity can be further shown by scrutinizing the expectations of the school.

THE EXPECTATION: "WHATEVER IS AMPLE INSIDE, TAKES SHAPE OUTSIDE."

The expectation of the school also showed that the literati's method of fostering artistic creativity was applied in a new way. As stated before, the understanding of the literati's artistic creativity had three basic elements. The third one, to seek creativity via the mental state (artistic moods, tastes, thoughts, and personality) rather than through technical representations (color, line, and style), was continued at Shanghai Art College, but it was then applied in a cross-cultural way. Similar to other artists of that era (for example, Xu Beihong and Lin Fengmian), artists at Shanghai Art College experimented with new methods of being creative by combining various components across oriental and occidental cultures. However, their solution, which was different from that of other artists in that era, was arguably (1) the combination of components from the two cultures at the spiritual level through deep and broad understanding; and (2) the use of either traditional literati or Western realistic artistic language for representation. However, they did not directly combine artistic components for representation (for example, using shading in Western art and lines in Chinese art together to create new artistic forms). Feng Zikai argued,

Future paintings will be harmonious combinations of Eastern and Western paintings. However, this combination is not half Western, half Chinese represented on papers, but combining the two painting methods in the artists' hearts, eyes and hands; not East plus West, but East multiplies West; not mixing the East and the West but *hua he* the East and West (achieving thorough understanding and a mastery over all the elements).[53] He Tianjian commented on "eclecticism" as follows:

They have sound reasons, trying to combine the East and the West would only result in non-East and non-West works (*bu zhong bu xi*). This thought is not wrong on the condition that the artist should have a solid grounding in both Western and Chinese studies.[54] Zheng Wuchang wrote:

In recent decades, the implantation of Western paintings provided an excellent opportunity for Chinese artists [to develop art]. However, without having even a superficial understanding of Western painting, they tried to reform Chinese painting and claimed that was creativity. This was a bias the same as an exaggeration about producing Chinese painting without studying Western painting. I believe Chinese painting and Western painting are not the same, and their painting techniques and the way of expression also differ from each other. It is not impossible for

Chinese painters to learn from the West but only those who have attained a certain level in Chinese paintings and are equally well grounded in Western painting are able to do that. Both Zeng Cun in the Ming period and Wu Li in the Qing period applied "Western" techniques to Chinese paintings, but the effects were not satisfactory. These cases suggest it is not an easy job. If only formal likeness (*xing si*) is pursued by using watercolor techniques to indicate the remote and the close or the concave and the convex, claiming that this is a combination of the East and West, this is purely meaningless.[55] Ni Yide argued that experiments with materials are completely dispensable. In other words, combining two artistic forms of representation is unnecessary:

Chinese and Western paintings should not be differentiated by the tools. If Westerners paint with a brush pen and *xuan* paper, they would still achieve Western paintings; vice versa, when Chinese artists paint with oil colors and canvas, they do Chinese paintings … when we borrow Western painting materials to represent the local colors, they are still Chinese paintings.[56] These Shanghai artists did not appreciate the efforts of their peers, such as the work of Gao Qifeng, Gao Jianfu, and other Lingnan School artists or even Xu Beihong and Lin Fengmian.[57] Thus a technical combination of Eastern and Western art was not favored; an ideal combination of East and West in their minds was capacity (basically a deep understanding) at the "mental state" and ideological level. They believed that appropriate new external techniques and styles to express ideas and finally to achieve the mixture of art in the two cultures existed only when these new scholarly thoughts were formed. As Fu Lei said, "I speak again! Where should we go? Go to the deep!"[58] Overall, the three standards of the literati's way of artistic creativity endured in art education at Shanghai Art College but were applied in a modern way. Students were encouraged to study diligently and deeply to attain artistic heights. The scholarship foundation and cultural background were broadened. Chinese and Western masters were imitated, and the cultural vision encompassed Western culture, nature, and society for breadth. As creativity at the spiritual level was pursued, the school urged students to deeply study Chinese and Western cultures instead of experimenting with artistic representational techniques to combine the two cultures. This pedagogy led to the growth of modern scholar–artists, who created new forms of art and expression with their wide view across disciplines and cultures, and modern scholar-style designers, whose designs were fundamentally different from that in Dadaism or Marcel Duchamp's

Fountain, Readymade. Shanghai Art College promoted artistic creativity by applying the literati's way of creativity in a new way.

NOTES

1. The author has written articles on art education in Republican Shanghai for the following journals: *Modern China, East Asian History, Modern Chinese Culture and Literature,* and *Studies in Art Education.*
2. Mayching Kao, "China's Response to the West in Art: 1898–1937" (PhD dissertation, Stanford University, 1972), 110.
3. Kao, "Conclusion" in "China's Response to the West in Art: 1898–1937."
4. According to Nikolans Pevsuer, hatred toward academies arose as early as the eighteenth century. Voltaire and the Encyclopedists questioned the value of academies. In the nineteenth century, Kleist tried to reform education by suggesting that young students invent instead of copy. There emerged more severe attacks on the unfavorable environment of academies for artists of genius. Whistler said, "The academy which the Gods wish to make ridiculous, they made Academicians." Fred Brown observed, "Throughout the school, every natural instinct of the student was prevented or frustrated." Nikolaus Pevsner, *Academies of Art Past and Present* (Cambridge: Cambridge University Press, 1940), 239.
5. "There is only ideal aim but no pedagogy in art education. … If there is pedagogy in art teaching and learning, it is only about two things: first, respecting personality (ge xing). As people differ at individual characters, they should never be enforced to be identical. … If this happens, their personal character will be suppressed and their artistic vitality is deprived. For this reason, the most important point in art education is students' individuality. The second important point is creativity. Creativity is an instinct of human beings. During its latent period (qian fu qi) and when it is occasionally touched, it becomes inspiration. Painting is the expression of thoughts and inspirations. … In conclusion, both art teaching and learning should be attached to the expression of creativity. Rigid imitation work never helps to reach the goal of art education." Liu Haisu, "How to Teach and Learn Painting," *Art Weekly* (Yishu zhoukan), no. 39 (1924); Zhu Jinlou and Yuan Zhihuang, eds., *The Selected Anthology of Liu Haisu's Art Essays* (Liu Haisu yishu wenxuan) (Shanghai: Shanghai People's Fine Arts Publishing House, 1987), 82–83.

6. Barry Keenan, *The Dewey Experiment in China: Educational Reforms and Political Power in the Early Republic* (Cambridge, MA: Harvard University Press, 1977), 61–62. Barry Keenan observed that in the May Fourth Movement and during educational reform in 1922, Dewey's lectures in China exerted important influence.
7. Keenan, *The Dewey Experiment*, 65.
8. Ibid., 66.
9. Michael Sullivan, *Twentieth Century Chinese Art and Artists* (Berkeley: University of California Press, 1996), 46.
10. Wang Geyi, *Random Reminiscences of Wang Geyi* (Wang Geyi Suixiang Lu) (Shanghai: Painting and Calligraphy Press, 1982), 48–49.
11. See record of the allied teaching and administrative staff meeting on October 15, 1931 (Q250-1-43), Archives of the Shanghai Art College.
12. Order from the Ministry of Education, no. 8530, forwarded by the Shanghai Education Bureau on July 23, 1934, in the correspondence documents between Shanghai Art College and the Ministry of Education (Q250-1-3), Archives of the Shanghai Art College.
13. See "Major Teaching Purposes for Each Discipline and Its Practice Situation" (Geke Jiaoshou Yaozhi ji qi Shishi Zhangkuang) in a survey of the Shanghai Art College in 1924, Archives of the Shanghai Art College.
14. Thomas Munro, "Creative Ability in Art and Its Educational Fostering," in *Art in American Life and Education*, ed. **Guy Montrose Whipple** (Chicago: University of Chicago Press, 1941), 289–321.
15. Wang Geyi, *Random Reminiscences*, 78.
16. See the record of the Normal Department meeting on March 17, 1924 (Q250-1-40), Archives of the Shanghai Art College.
17. See, for example, the record of the teaching affairs meeting on April 15, 1924 (Q250-1-40), Archives of the Shanghai Art College.
18. Liu Haisu, "Looking Back at the Ten Years of the Shanghai Art School" (Shanghai Meizhuan shinian huigu), in *A Selection of Liu Haisu's Art Essays* (Liu Haisu Yishu Wenxuan), ed. Zhu Jinlou and Yuan Zhihuang (Shanghai: People's Press, 1987), 36–42. The score system resumed in the 1930s, primarily because of regulations of the Education Ministry.

19. Cai had a habit of randomly transforming the appearance and body shape of models according to his own aesthetic standards. Both teachers Wang Yuanbo and Fan Xinqiong did not appreciate his style. They did not criticize his paintings or force him to change this habit, but they also did not comment on his paintings. See Cai Ruohong, *The Fashion at the Pavilion's Era in Shanghai* (Shanghai Tingzijian Shidai de xiqi) (Shijiazhuang: Hebei Education Press, 1999), 49–50.
20. Matthew Arnold, *Culture and Anarchy* (London: Smith, Elder & Co., 1882), www.gutenberg.org/ebooks/4212 (accessed June 6, 2006).
21. Creativity in literati painting is interpreted by Sullivan as something similar to creativity in music. See Sullivan, *Twentieth Century Chinese Art and Artists*.
22. Li Mengyang, one of the *hou qi zi* in the Ming period, pointed out, "In the creations of the ancient master builders … every hall was different and every door was not alike; yet in executing the square and the circle, no one could avoid using the ruler and the compass. … The ruler and the compass represent the methods (fa). … If someone can do without the ruler and the compass, he must still be able to execute the square and the circle, then he may forget about [the ruler and the compass]." Cited in Wen Fong, *Images of the Mind* (Princeton, NJ: Princeton University Art Museum, 1984), 157.
23. Dong Qichang stated, "Painting must be formed using ancient methods. Whenever the brush is applied to paper, every stroke shows its source and tradition." Cited in Wai-kan Ho, "Tung Ch'i-ch'ang's Transcendence of History and Art," in *The Century of Tung Ch'i-ch'ang 1555–1636,* vol.1, ed. Wen Fong (Kansas City, MO: Nelson-Atkins Museum of Art; Seattle: University of Washington Press, 1993), 3–40.
24. Susan Bush, *The Chinese Literati on Painting Su Shib (1037–1101) to Tung Ch'i-ch'ang (1555–1636)* (Cambridge, MA: Harvard University Press, 1971), 13.
25. Ibid., 7.
26. Lu Yanshao, *Reminiscences of Lu Yanshao* (Lu Yanshao Zixu) (Shanghai: Painting and Calligraphy Press, 1986).
27. Bush, *Chinese Literati*, 184.
28. For a translation, see Bush, *Chinese Literati,* 145.

29. See "Major Teaching Purposes for Each Discipline and Its Practice Situation" (Geke Jiaoshou Yaozhi ji qi Shishi Zhangkuang) in survey of the Shanghai Art College in 1924, , Archives of the Shanghai Art College.30. "Comments on Li Xiuqiong," *The Seventeenth Graduation Book of Shanghai Art College* (Shanghai: Shanghai Art College, 1936).
30. Si Du, "Biography of Yang Xinqi," *The Sixteenth Graduation Book of Shanghai Art College* (Shanghai: Shanghai Art College, 1935).
31. Ibid., 72.
32. Zhi Wu, "Biography of Cheng Yaozeng," *The Fifteenth Graduation Book of Shanghai Art College* (Shanghai: Shanghai Art College, 1934), 70.
33. At this point, Shanghai Art College was not a special example but a reflection of the art field of the whole academic atmosphere in China in that era. Working hard was also emphasized in Japanese traditional art training. See John Singleton, ed., *Learning in Likely Places: Varieties of Apprenticeship in Japan* (Cambridge: Cambridge University Press, 1998).
34. For compulsory courses, see the second section in this chapter.
35. When the Chinese Painting Department was first established, it showed great influence from the Beijing Art School, where Chinese painting–related subjects, such as poetry, calligraphy, and literature, were excluded. Most courses were Western painting courses. This was soon reformed at Shanghai Art School. The schedule in 1928 showed a rational arrangement, with compulsory Chinese painting and elective Western painting courses.
36. See the regulations of the Shanghai Art College in June 1937, Archives of the Shanghai Art College.
37. Ibid.
38. See the school survey in 1925 and 1937, Archives of the Shanghai Art College. At Shanghai Art College, applied art education aimed to create modern designers who could think and study instead of artisans, which made applied art education at Shanghai Art College essentially different from that in India. The purpose of art schools in the latter was to create artisans. See Partha Mitter, *Art and Nationalism in Colonial India: 1580–1922* (New York: Cambridge University Press, 1994), 36.

39. Liu Haisu, *A New Impression of Japanese New Art* (ri ben mei shu de xin ying xiang) (Shanghai: Commercial Press, 1921); Zhu Jinlou and Yuan Zhihuang, *A Selection of Liu Haisu's Art Essays,* 205.
40. Lu Yifei, "On 'three understandings' in Memory of Mr. Gai Jiaotian," in *A Commemorative Collection of LuYifei* (Lu Yifei Jinian Wenji), ed. Shen Zuan (Gu Wuxuan Press, 2002), 8–11.
41. Dong Qichang wrote, "A painter must pass two critical tests. First, he must imitate the ancient masters; then he must imitate nature"; "The painter who models himself after the ancient masters already belongs to the Upper Vehicle. Advancing one more step, he must make himself after heaven and earth." Dong Qichang, *The Eye of Painting* (hua yan), vol. 12 (Taibei: World Press, 1962), 27.
42. Wen Fong, "Tung Chi-chang and Artistic Renewal," in *The Century of Tung Ch'i-ch'ang 1555–1636*, vol. 1, ed. Wen Fong (Seattle: University of Washington Press, 1993), 43–54.
43. Xu Lantai, a student and foster son of Wu Fufan, told me that he once asked Wu what the reason was. Wu said that most mountains and water are quite similar. Wu was proud of his room full of students and visitors every day and called himself the scholar (*xiu cai*) who knows all the world's affairs without having to step outside his gate. Private interview with Xu Lantai, Shanghai, December 22, 2001.
44. Record of a school meeting on March 30, 1925 (Q250-1-40), Archives of the Shanghai Art College.
45. Record of a Chinese Painting Department meeting on September 20, 1926 (Q250-1-41), Archives of the Shanghai Art College.
46. Record of a school meeting on April 12, 1935 (Q250-1-32) , Archives of the Shanghai Art College.
47. Yu Jianhua, "Life Drawing in Chinese Painting" (guo hua xie sheng), in *A Selected Anthology of Yu Jianhua's Art Essays* (Yu Jianhua mei shu lun wen xuan), ed. Zhou Jiyan (Jinan: Shandong Fine Arts Publishing House, 1986), 106–108.
48. Ibid.
49. On March 19, 1920, Wang Jiyuan suggested a monthly student art competition (Q250-1-35).
50. *Shanghai Daily*, May 13, 1923, and May 7, 1933.
51. *Shanghai Daily*, January 12, 1937.

52. Feng Zikai, "Future Paintings" (jiang lai de hui hua), in *Random Essays on Art* (Yishu Cong Hua) (Hong Kong: New Literature Research Institution, 1976), 60.
53. He Tianjian, "My Opinions on Chinese Paintings" (Wo duiyu Guohua zhi Zhuzhang), *Art and Life* 1, no. 3 (1934). In this article, he also proposed replacing Chinese paintings with Western paintings to reform Chinese painting mixing together Chinese paintings with Western paintings. He criticized people who held these opinions as being stimulated by incomplete emotions without rational consideration.
54. Zheng Wuchang, "A Reply to Wen Bingdun" (Da Wen Bingdun Xuedi), *Shanghai Art College Quarterly*, no. 2 (1929): 50.
55. Ni Yide, "From Landscapes to Traveling Life Paintings" (cong feng jing hua shuo dao lu xing xie sheng), *The Fourteenth Graduation Book* (Shanghai: Shanghai Art College 1934), 139.
56. Gao Jianfu combined light and shading in Western art and Chinese brush-ink techniques in his works. Xu Beihong used the nude in Western art to represent Chinese historical stories and combined shading and lines.
57. Fu Lei, "I say again, where should we go? Go to the deep!" *Arts* (yi shu), no. 2 (1933): 11–13. This article is one in a discussion about "In which direction should Chinese art develop?" In this article, Fu Lei suggested that people could decide only after a thorough understanding of Eastern and Western artistic thought and theories.

Chapter 5

Combating Cultural Heritage Crimes: Recent Developments in China

Minxing Zhao

Over the past several decades, increased recognition of the criminal threats to cultural heritage has prompted intensified criminal justice response to these crimes in China. Since the enactment of the first criminal law in 1979, a total of eleven heritage offenses with stringent punishments have been added to the penal code through amendments to the law and judicial interpretations. In the meantime, China works to harness the potential of the 1970 UNESCO Convention on the Means of Prohibiting and Preventing the Illicit Import, Export and Transfer of Ownership of Cultural Property to establish law enforcement cooperation structures with other countries to further thwart trafficking in Chinese cultural heritage.

This chapter examines the efforts China has made to fight heritage crimes in recent years. In particular, this analysis focuses on recent developments in Chinese criminal law and bilateral agreements with other countries. The impact these developments may have on China's long-term and sustainable capacity building in the protection of cultural heritage is also explored. The first section briefly introduces the application and removal of the death penalty for heritage crimes and discusses the implications of recent penal reform. The second section reviews the new judicial interpretation of criminal law concerning the theft of cultural relics. The third section offers an overview of developments in China's bilateral cooperation with other countries, followed by a brief conclusion.

THE ABOLITION OF THE DEATH PENALTY FOR CULTURAL HERITAGE CRIMES

Although China habitually relies on the death penalty to maintain social order and societal control, the use of the death penalty to punish cultural heritage crimes lasted for only a short period, from the 1980s to 2011. In 1979, when China adopted its first criminal law, there were twenty-eight capital crimes, and the death penalty was applied to overtly political crimes.[1] Illicit transportation of precious cultural relics was the only heritage offense at that time, and the highest punishment for this crime was life imprisonment, not capital punishment.[2] In a major revision made to criminal law in 1997, however, the number of capital offenses greatly expanded, from twenty-eight to sixty-eight, in response to a surging crime wave and the emergence of new criminal activities during the course of economic reforms after the 1980s.[3] Four capital offenses—the smuggling of cultural relics, theft of valuable cultural relics, illegal excavating and robbing of ancient cultural sites or ancient tombs, and illegal excavating and robbing of hominid fossils or vertebrate fossils—were incorporated into the criminal law.[4] After 1997, the sixty-eight capital offenses remained stable until a 2011 amendment to the criminal law removed the death penalty for thirteen economic-related nonviolent crimes, including all cultural heritage crimes.[5]

The abolition of the death penalty for heritage crimes has different implications for China's capital punishment reform and the protection of cultural heritage. In 2010 the National People's Congress (NPC) of China released the draft amendment to the criminal law to the public to solicit comments and opinions. In explanation of the draft amendment, the NPC made it clear that the reason for abolishing the thirteen capital offenses was that the death penalty was seldom or never applied for those crimes.[6] Thus, for the legislature, keeping the death penalty for heritage crimes and other economic-related nonviolent crimes was only of symbolic value, without any practical significance. On the other hand, the removal of capital punishment for these crimes significantly reduced the number of crimes carrying the death penalty and put criminal law reform in line with the policy of "killing fewer, killing cautiously" and "balancing leniency and severity" in the course of constructing a harmonious society.[7]

This paper argues that for the protection of cultural heritage, it is more important to match efforts to combat crimes against cultural heritage with either the gravity or the extent of crimes than to simply rely on the deterrence provided by the death penalty. Over the years, the fact that the death penalty was rarely used certainly did not mean that there were no serious crimes against cultural heritage. The substantial amount of ancient artifacts looted and trafficked to international markets indicates that the death penalty did not seem to have the expected deterrence effect. The abolition of the death penalty for cultural heritage crimes, although not very welcomed among practitioners in the cultural heritage preservation and conservation field, may not increase the threat to cultural heritage as long as consistent and strengthened law enforcement efforts are maintained to counter these crimes.

NEW JUDICIAL INTERPRETATION CONCERNING THE THEFT OF CULTURAL RELICS

Apart from the abolition of the death penalty for all cultural heritage crimes in 2011, the new judicial interpretation concerning the theft of cultural relics was another major change made to the criminal justice response to heritage crimes. On April 3, 2013, The Supreme People's Court and the Supreme People's Procuratorate jointly issued "Interpretation of the Supreme People's Court and the Supreme People's Procuratorate on Several Issues Concerning the Application of Law in the Handling of Criminal Cases of Theft (Judicial Interpretation)." Building on the previous amendment to criminal law in 2011, this judicial interpretation clarified vague provisions in the criminal law with respect to the threshold of the theft of cultural relics and other related issues.[8]

First, article 9 of the judicial interpretation clarified the threshold requirements for criminal prosecution of cultural relics theft. In the amendment to criminal law in 2011, article 264 sets the threshold for conviction for all theft crimes based solely on the monetary value of the stolen property, without taking into consideration the special characteristics of the theft of cultural relics. Article 264 states,

Whoever steals a relatively large amount of public or private property, commits thefts many times, commits a burglary or carries a lethal weapon to steal or pick pockets shall be sentenced to imprisonment of not more than 3 years, criminal detention or control and/or a fine; if the amount involved is huge or there is any other serious circumstance, shall be sentenced to imprisonment of not less than 3 years but not more than 10 years and a fine; or if the amount involved is especially huge or there is any other especially serious circumstance, shall be sentenced to imprisonment of not less than 10 years or life imprisonment and a fine or forfeiture of property.[9]

The above provision uses vague terms, such as *relatively large, huge,* and *especially huge* to describe the amounts required to meet the criminal threshold for the prosecution of theft crime. For ordinary property theft crimes, the judicial interpretation translates these vague standards into readily identifiable monetary terms. According to article 1 of the judicial interpretation,

Whoever steals public or private properties of a value of between 1,000[10] and 3,000 or more, 30,000 and 100,000 or more, and 300,000 and 500,000 or more shall be respectively determined as having stolen properties of a "relatively large amount," "huge" and "especially huge" as prescribed in article 264 of the criminal law.[11]

Due to a lack of recognized and widely accepted methodologies for the assessment of the economic values of stolen cultural heritage property, the standards for ordinary property theft cannot be applied to punish heritage crimes. To rectify this legislative deficiency, article 9 of the judicial interpretation uses the grading system from the cultural relics protection law to determine standards for punishing heritage crimes. According to article 3 of the Law of the People's Republic of China on Protection of Cultural Relics,

> Movable cultural relics, such as important material objects, works of art, documents, manuscripts, books, materials, and typical material objects dating from various historical periods, shall be divided into valuable cultural relics and ordinary cultural relics; and the valuable cultural relics shall be subdivided into grade-one cultural relics, grade-two cultural relics and grade-three cultural relics.[12]

The first section of article 9 of judicial interpretation explains,

Whoever steals ordinary cultural relics, grade-three cultural relics, grade-two cultural relics or above owned by the state and collected in state museums shall be respectively determined as having stolen properties of a "relatively large amount," "huge" and "especially huge" as prescribed in Article 264 of the Criminal Law.[13]

Section 2 of article 9 stipulates that if the theft involves multiple cultural relics owned by the state and collected in state museums, three same-grade cultural relics can be regarded as one higher-grade cultural relic. For example, if a case involves the theft of three ordinary cultural relics and two grade-three cultural relics, those three ordinary cultural relics can be regarded as one grade-three cultural relic. Then this new grade-three cultural relic and the other two grade-three cultural relics can be regarded as one grade-two cultural relic. After this upgrading process, the amount of theft reaches the "especially huge" level, and the highest penalty for this case may be lifelong imprisonment.

For the theft of privately owned cultural relics, their value can be determined by market price. If there is no price for such cultural relics or it is unreasonable to determine the value of stolen property based on price, article 9 states that the value of the stolen cultural relics can be assessed by price evaluation agencies. Here it is obvious that the judicial interpretation makes a distinction between state-owned and privately owned cultural relics when assessing their values. Although this distinction may lead to unequal protection of state-owned and privately owned cultural relics, the provision discussed here actually reflects and responds to the current situation of cultural heritage protection in China. In China, according the constitution and the Cultural Relics Protection Law, state ownership of cultural heritage is the rule and private ownership of cultural relics is the exception.[14] State-owned cultural relics are usually hoarded or displayed in state museums. This makes state-owned cultural relics easier targets than privately owned cultural relics; accordingly, state-owned relics require more legal protection.

Punishment for attempted theft of valuable cultural relics depends on the gravity of the act. Since valuable cultural relics are divided into three grades, the attempted theft of grade-three, grade-two, or grade-one relics is punished according to judicial interpretation. Attempted theft of valuable cultural relics, though not necessarily resulting in the loss of cultural property, represents a grave threat to cultural heritage. In addition, punishing attempted theft can deter potential heritage theft.

BILATERAL AGREEMENTS ON THE PROTECTION OF CHINESE CULTURAL PROPERTY

As a signatory to the 1970 UNESCO Convention on the Means of Prohibiting and Preventing the Illicit Import, Export and Transfer of Ownership of Cultural Property, China has been active in entering bilateral agreements with other member states on cooperation to prevent the theft, clandestine excavation, and illicit import and export of cultural property. Since the first bilateral agreement was concluded in 2000 with Peru, China has reached such agreements or memorandums of understanding with eighteen countries around the world as of this writing (see Table 5.1). As indicated in Table 5.1, although the process of negotiation was not easy, recent years have witnessed China's intensified steps in soliciting support from member states for the 1970 UNESCO convention for its efforts devoted to cultural heritage protection.

Table 5.1. Countries entered into bilateral agreements with China (in chronological order)

Country	Year
Peru	2000
Italy	2006
India	2006
Philippines	2007
Greece	2008
Chile	2008
Venezuela	2008
United States	2009
Turkey	2009
Ethiopia	2009
Australia	2009
Egypt	2010
Mongolia	2011
Mexico	2012
Colombia	2012
Nigeria	2013
Switzerland	2013
Cyprus	2013

Source: Ministry of Foreign Affairs of the People's Republic of China, http://www.fmprc.gov.cn/mfa_chn/ziliao_611306/tytj_611312/

Despite the progress China has made in enlisting international support for its endeavor to strengthen cultural heritage security, the inherent shortcomings of the current way of concluding bilateral agreements and the aspirational nature of the agreements fail to demonstrate that China has clear strategies to reduce threats to its cultural heritage from international criminal activities.

First, a majority of these agreements were signed with cultural-heritage-rich source countries rather than with cultural-heritage-poor market countries. For the past several decades, the rising demand for cultural heritage in the Western world—that is, market countries—has resulted in the enormous outflow of ancient artifacts, legal and illegal, from source countries such as China. Therefore, agreements with source countries will not effectively stem the flow of cultural heritage from China to market countries. A memorandum of understanding signed with one of the most important market countries, the United States, on January 14, 2009, was a breakthrough in seeking international support, but the practical value of the agreement cannot be underestimated. The focus of the memorandum is more on how China should increase efforts to prevent the pillage of cultural heritage material and its entry into illegal markets than on how the United States should implement import restrictions on specified categories of archaeological material originating in China.[15] As with the agreements with source countries, the symbolic meaning far exceeds true practical value. In the agreement with the United States, China was urged to "strengthen regional cooperation within Asia for the protection of cultural patrimony" and to "seek increased cooperation from other importing nations to restrict the import of looted archaeological material originating in China."[16] Nevertheless, these agreements at least show the world that China is determined to combat heritage crimes. The moral support gained from other source countries definitely consolidates China's bargaining position with market countries.

Second, China has not entered into many agreements with neighboring countries or territories. The international trafficking in Chinese cultural heritage usually uses neighboring countries or territories as transit points. The agreements with neighbors will greatly reduce the incentive for pillage of China's rich cultural heritage. Without such agreements, cooperation with Western market countries may not achieve the desired goals. In the Chinese–US memorandum of understanding, China is also reminded to "make every effort to stop archaeological material looted or stolen from the Mainland from entering the Hong Kong Special Administrative Region and the Macao Special Administrative Region with the goal of eliminating the illicit trade in these regions."[17]

Third, most of these agreements are aspirational in nature, without any operational capabilities and with no consequences for noncompliance. Since most of the agreements were signed with source countries rather than with market countries, it is hard for source countries to formulate enforceable measures to deter trafficking crimes, as most of them are not destinations for Chinese cultural relics. For the few agreements with market countries such as the United States, due to divergent attitudes toward cultural heritage, it is even less likely for concrete law enforcement measures to be included.

CONCLUDING OBSERVATIONS

As a country endowed with rich cultural heritage, China is serious about the protection of its historical and cultural legacies. As discussed in this chapter, concerted effort has been made to enhance domestic criminal legislation and to tighten international law enforcement cooperation. Despite progress over the past few decades, in terms of both criminal legislation and international law enforcement measures, there are still enormous gaps between the goals of crime control and the actual extent or gravity of heritage crimes. The rising demand for cultural property in domestic and international markets will make efforts to counter crimes against cultural heritage more challenging. Accordingly, future criminal justice measures need to be more responsive to new developments in criminal trends.

NOTES

1. See Dingjian Cai, "China's Major Reform in Criminal Law," *Columbia Journal of Asian Law* 11 (1997). Also see Jeremy Monthy, "Internal Perspectives on Chinese Human Rights Reforms: The Death Penalty in the PRC," *Texas International Law Journal* 33 (1998).
2. According to article 173 of the 1979 criminal law (not now in force), "Violation of the cultural relic protection regulations by the illicit transportation of precious cultural relics is punishable by fixed term imprisonment of not fewer than 3 years but not more than 10 years, and also by fines. Where the circumstances are 'serious,' the offense is punishable by fixed-term imprisonment of not fewer than 10 years or by life imprisonment, and also by confiscation of property." The article is available at http://www.novexcn.com/criminal_law.html.
3. See Hong Lu and Lening Zhang, "Death Penalty in China: The Law and the Practice," *Journal of Criminal Justice* 33 (2005).
4. For a detailed history of China's death penalty, see Hong Lu and Terance D. Miethe, *China's Death Penalty: History, Law, and Contemporary Practices* (New York: Routledge, 2007), 50–56.
5. Amendment 8, adopted by the Nineteenth Meeting of the Standing Committee of the National People's Congress on February 25, 2011, and effective on May 1, 2011, is available at http://www.npc.gov.cn/huiyi/lfzt/xfxza8/2011-05/10/content_1666059.htm (in Chinese) and http://www.cecc.gov/resources/legal-provisions/eighth-amendment-to-the-criminal-law-of-the-peoples-republic-of-china (in English translation).
6. For details, see the draft amendment to the criminal law and accompanying explanations at http://www.npc.gov.cn/npc/flcazqyj/2010-08/28/content_1592773.htm.
7. See Susan Trevaskes, "The Death Penalty in China Today: Kill Fewer, Kill Cautiously," *Asian Survey* 48, no. 3 (2008), doi 10.1525/as.2008.48.3.393.
8. For more information on the issuance and application of judicial interpretation, see Ronald C. Keith and Zhiqiu Lin, "Judicial Interpretation of China's Supreme People's Court as 'Secondary Law' with Special Reference to Criminal Law," *China Information* 23, no. 2 (2009), doi 10.1177/0920203X09105126.
9. See note 5.
10. In Chinese yuan.

11. Available at http://www.chinacourt.org/law/detail/2013/04/id/146160.shtml (in Chinese) and http://hk.lexiscn.com/law/interpretation-of-the-supreme-peoples-court-and-the-supreme-peoples-procuratorate-on-several-issues-concerning-the-application-of-law-in-the-handling-of-criminal-cases-of-theft.html (in English translation).
12. The Law on the Protection of Cultural Relics is available at http://www.sach.gov.cn/art/2007/10/29/art_1591_58336.html.
13. The text is available in Chinese at http://www.chinacourt.org/law/detail/2013/04/id/146160.shtml. The English translation is provided by the author.
14. See article 22 of the Constitution of the People's Republic of China, available at http://www.gov.cn/gongbao/content/2004/content_62714.htm. Also see article 5 and article 6 of the Law on the Protection of Cultural Relics, available at http://www.sach.gov.cn/art/2007/10/29/art_1591_58336.html.
15. Rhonda Schechter, "Preventing Pillage and Promoting Politics: The Dual Goals of the 2009 U.S.–China Bilateral Agreement to Restrict Imports of Chinese Cultural Property," *Art, Antiquity and Law* 14, no. 4 (2009).
16. Article 2 of the Memorandum of Understanding between the Government of the United States of America and the Government of the People's Republic of China Concerning the Imposition of Import Restrictions on Categories of Archaeological Material from the Paleolithic Period through the Tang Dynasty and Monumental Sculpture and Wall Art at Least 250 Years Old is available at http://eca.state.gov/files/bureau/ch2009mou.pdf.
17. Ibid.

BIBLIOGRAPHY

Cai, Dingjian. "China's Major Reform in Criminal Law." *Columbia Journal of Asian Law* 11 (1997): 213–18.

Keith, Ronald C., and Zhiqiu Lin. "Judicial Interpretation of China's Supreme People's Court as 'Secondary Law' with Special Reference to Criminal Law." *China Information* 23, no. 2 (2009): 223–55. doi 10.1177/0920203X09105126.

Lu, Hong, and Lening Zhang. "Death Penalty in China: The Law and the Practice." *Journal of Criminal Justice* 33 (2005): 367–76.

Lu, Hong, and Terance D. Miethe. *China's Death Penalty: History, Law, and Contemporary Practices.* New York: Routledge, 2007.

Monthy, Jeremy. "Internal Perspectives on Chinese Human Rights Reforms: The Death Penalty in the PRC." *Texas International Law Journal* 33 (1998): 189–226.

Schechter Rhonda. "Preventing Pillage and Promoting Politics: The Dual Goals of the 2009 U.S.–China Bilateral Agreement to Restrict Imports of Chinese Cultural Property." *Art, Antiquity and Law* 14, no. 4 (2009): 317–47.

Trevaskes, Susan. "The Death Penalty in China Today: Kill Fewer, Kill Cautiously." *Asian Survey* 48, no. 3 (2008): 393–413.

Section 2: Hong Kong

Chapter 6

The Role of Hong Kong Government in Creative Industries: Analyzing the Cultural Discourse of Three Policy Addresses

Ming Lai and Chi Cheung Leung

The chief executive (CE) of the Hong Kong Special Administrative Region (HKSAR) delivers a policy address to Legislative Council members and the people of Hong Kong every year. From 1997 to 2013, a total of sixteen policy addresses were delivered by three CEs, namely Tung Chee-hwa, Donald Tsang, and Leung Chun-ying. The policy addresses serve to express the policy visions, governance, and major concerns of HKSAR leaders during their terms of office. The purpose of this chapter is to analyze the cultural discourse expressed in policy addresses since the establishment of the HKSAR. Compared with such issues as economic development, education, and people's livelihood, the proportion of culture in policy addresses is undoubtedly far less. Hong Kong does not have a clear cultural policy. Even with only a handful of sentences, the way culture has been articulated in policy addresses would be an invaluable reference in evaluating government policies. This study analyzes the policy addresses in the areas of arts, culture, and creative industries. It traces transformations in policy foci of the government and discerns the styles of governance of the three leaders. Based on the concept of keyword and thematic analyses, this chapter reviews the cultural discourse of the CEs. By measuring the frequency of the appearance of culture-related keywords, it reveals how culture was and is treated by the administration. In addition, the chapter summarizes themes of cultural discourse in the policy addresses and delineates the differences and similarities between the three CEs' policies.

KEYWORD ANALYSIS

This section presents analysis of the appearance of culture-related keywords from the policy addresses of the three CEs of the HKSAR, which are accessible through government websites. The first keyword identified is culture (文化 wen hua). Arts (藝術 yi shu), which bears a close relationship with culture, is identified as another keyword. In more recent years, culture is also linked to the terms creativity (創意 chuang yi) and industries (產業 chan ye). The establishment of the West Kowloon Cultural District (WKCD; 西九文化區 xi jiu wen hua qu) is an important event in the cultural scene in Hong Kong. Hence the keywords creativity, industries, and West Kowloon are also invoked in this analysis.

The keyword analysis is based on searches for Chinese words in the policy addresses, originally delivered in Chinese by the CEs. A search conducted in Chinese effectively eliminates the problem of variation, as words such as *wen hua* and *chuang yi* can be used as both nouns and adjectives, and words such as *chan ye* bear no distinction between singular and plural forms. Versions in English are available on government websites, but those who search in English need to aware of variants of the same terms, such as *culture/cultural* and *industry/industries*. By counting the appearance of these culture-related keywords in policy addresses, this study reasonably assesses how the discourse of culture in policy has changed over the years.

Table 6.1 Frequency of appearance of culture-related keywords in policy addresses of chief executives in past years

Chief Executive	Month/ Year	Frequency of Appearance of Keyword				
		culture (*wen hua*)	arts (*yi shu*)	creative (*chuang yi*)	industries (*chan ye*, with *creative* or *cultural* as the prefix)	West Kowloon (*xi jiu*, excluding traffic issues)
C. H. Tung (First term)	10/1997	27	10	8	0	0
	10/1998	19	15	2	0	1
	10/1999	23	0	4	0	1
	10/2000	18	0	3	0	0

	10/2001	10	1	4	1	1
C. H. Tung (Second term)	1/2003	5	4	9	6	0
	1/2004	3	1	7	3	0
	1/2005	28	4	25	22	2
Donald Tsang (First term)	10/2005	19	8	8	3	4
	10/2006	16	10	8	6	1
Donald Tsang (Second term)	10/2007	42	18	20	7	8
	10/2008	14	4	9	8	3
	10/2009	24	6	15	10	4
	10/2010	12	7	2	1	3
	10/2011	11	7	4	3	1
C. Y. Leung	1/2013	34	23	15	8	4

As shown in Table 6.1, the word *culture* appears in double digits throughout the policy addresses, except in 2003 and 2004, when the number of appearances dropped drastically to five and three, respectively. In contrast, the words *creative* and *industries* appeared more frequently in those two years. HKSAR leaders used *culture* less and *creative* and *industries* more. From 2005 onward, appearances of all three keywords surged. In the 2005 policy address, Tung employed *cultural and creative industries* as an official term, and all three keywords appeared more than twenty times in 2005. But in Tung's first two years in office, 1997 and 1998, the word *arts* appeared ten times or more. However, in the following two years, 1999 and 2000, *arts* totally disappeared from policy addresses. Interestingly, with the rise of discourse about industries and West Kowloon in recent years, especially since Donald Tsang assumed the duty of CE, the word *arts* has become more frequent in the discourse of HKSAR leaders.

In January 2013, C. Y. Leung delivered his first policy address, which he described as a policy blueprint for the next five years. One characteristic of this address was that frequencies of all culture-related keywords were high. *Culture* appeared thirty-four times, and *arts* appeared twenty-three times—the most appearances among all policy addresses. Such a pattern was comparable to the first policy address delivered by Donald Tsang in his second term (2007), in which the frequencies of all the keywords were high. Tsang said *culture* forty-two times. Like Leung, he regarded his 2007 policy address as a policy blueprint for the next five years. Both of them mentioned a variety of things related to culture in the first policy addresses of their new five-year terms of office.

THEMATIC ANALYSIS

Keyword analysis provides an assessment of how the cultural discourse of HKSAR leaders changes through the years. Following this, a thematic analysis of culture-related paragraphs in policy addresses supports a deeper understanding of the contents of cultural discourse. This section presents the analysis, beginning with findings related to Tung. Unless otherwise stated, the direct quotes in the following paragraphs are based on English translations of policy addresses published on government websites.

C. H. Tung's Era

A total of eight policy addresses were delivered by C. H. Tung from 1997 to his resignation in 2005. Throughout the years, two major viewpoints could be identified in the cultural discourse expressed in his policy addresses. The first was that Hong Kong, being a place where Eastern culture meets Western culture, could play an important role in bridging China and the Western world. Another view was that Hong Kong people should learn more about the culture of the motherland. The following paragraph is found in his first policy address, delivered in 1997:

> Ours is a cosmopolitan city. Our ability to embrace the cultures of east and west is one of the secrets of our success. ... While we deepen our understanding of Chinese history and culture, we will continue to develop our own diverse cultural characteristics. China's culture ... is growing and changing as we journey forward into the 21st Century. Hong Kong stands in a unique position in this process, able to act as the center of exchange for China to learn about western cultures and for the world to learn about Chinese culture (Tung 1997).

These two major viewpoints keep appearing in Tung's policy addresses—for example, the one delivered in 1999:

> Our reunification with the motherland has enabled us to build on Chinese culture and at the same time draw on Western culture to develop our own distinctive and colourful culture. We will continue to promote public understanding of Chinese culture, history and heritage on the one hand and to enhance our exchanges and communication with the rest of the world on the other, so as to learn from different cultures around the world. ... I have proposed to develop Hong Kong into an international centre for cultural exchanges. This will help to strengthen our identity as a world-class city (Tung 1999).

On the other hand, in the years 1997 and 1998, there were paragraphs specifically dedicated to arts, such as the following:

> Young and old alike look to our city to be far more than just a place of study and business. We look for art to stimulate and sustain us. Hong Kong has long embraced both eastern and western cultures and in our artistic life we find contemporary diversity with Chinese characteristics. Over the years we have injected considerable resources into developing artistic endeavors. ... My Administration will consider how to make better use of the resources we now invest in this area, and what more we can do to stimulate the artistic life of Hong Kong (Tung 1997).

The embrace of Eastern and Western cultures was also mentioned in this paragraph. In addition to the two major viewpoints, new elements were added to the policy addresses through the years. In 1999 Tung mentioned the relationship between culture and creativity: "What we also need is a favorable and flourishing cultural environment that is conducive to encouraging innovation and creativity in our citizens" (Tung 1999). In 2000 he turned to the changes brought about by the "knowledge economy," as well as the impacts of creativity and technology on economic development:

> In a knowledge-based economy, anyone equipped with knowledge and creativity stands a chance of succeeding regardless of his or her social status or family background. ... Three years ago, there were very few people in Hong Kong who thought about the relationship between technology

and economic development. Now it is widely recognized that innovation and technology are essential to enhance productivity for our sustained economic growth (Tung 2000).

The year 2000 also marked the establishment of the government Leisure and Cultural Services Department (LCSD), replacing the Urban Council and Regional Urban Council regarding the organization and implementation of cultural matters. Tung stated,

> Culture and Heritage Commission and Leisure and Cultural Services Department are working with vision to instill a passion for learning and intellectual pursuits, arouse greater civic consciousness, environmental awareness, and concern for professional ethics and conduct [and enhance arts appreciation among the public, so that the culture of Hong Kong can be enriched comprehensively] (Tung 2000).

The bracketed phrase in the above paragraph, which was not translated in the English version, was added by the authors based on the original Chinese version of the policy address. In 2002 C. H. Tung was reelected as CE. Besides reiterating "As the place where East meets West, we believe that the interplay of different cultures will stimulate diversity and inspire creativity" (Tung 2003), this policy address articulated "creative industries" in more detail:

> Creative industries are important elements of a knowledge-based economy. The knowledge and wisdom of Hong Kong people, their innovative entrepreneurial spirit and agility combine to form a sound foundation. ... They [creative industries] are the synergy of artistic creativity and product development. ... The Secretary for Home Affairs, the Secretary for Commerce, Industry and Technology, and relevant bureaus and departments will work together to devise a concrete plan and create the necessary favorable environment to promote and facilitate the development of these creative industries (Tung 2003).

In 2004 Tung's major viewpoint of Hong Kong being a bridge between Eastern and Western cultures was linked to the "creative industries." He mentioned that the film industry had benefited by establishment of the

Closer Economic Partnership Arrangement (CEPA) between Hong Kong and mainland China. He remarked:

> For many decades, Hong Kong has been a place where the cultures of East and West meet. This is conducive to the development of creative industries . . . the Commerce, Industry and Technology Bureau and the Home Affairs Bureau will promote the development of creative industries, including their linkage with the resources and markets in the Mainland so that they can reach new heights. Following CEPA, local films can be released in the Mainland without import quota restrictions starting from this year. This provides an unprecedented opportunity for our film industry to prosper (Tung 2004).

In January 2005 Tung delivered his last policy address. He used seven paragraphs, under the title "Developing Cultural and Creative Industries," to delineate various aspects of these industries, including the successful experience of the United Kingdom in creative industries, the advantageous position of Hong Kong, the categories of the industries, the need for setting up a consultative framework, and official use of the term *cultural and creative industries* to replace *creative industries*. Under the title "Urban Renewal" was a paragraph describing how, by renewing an old district, a cultural atmosphere could be created to foster cultural and creative industries. This explains why the culture-related keywords appeared frequently in 2005, as shown in Table 6.1.

Donald Tsang's Era

In March 2005 Tung resigned. Donald Tsang, the former chief secretary, took up the position as acting CE and then CE of the HKSAR. In October 2005 Tsang delivered his first policy address, while his last policy address was delivered in October 2011. In these seven policy addresses, the cultural discourse expressed can be summarized as two pillars plus many concrete projects. The two pillars were the West Kowloon Cultural District and cultural and creative industries. The following two paragraphs are found in his first policy address:

> We have conducted a six-month public consultation on the development of an integrated cultural district in West Kowloon. The results show that the community generally supports the development of a cultural landmark in West

Kowloon. They consider that it will not only enrich our cultural and arts life, but also promote tourism and create jobs (Tsang 2005).

Through the Commission on Strategic Development, we will explore practical measures, including creating an enabling environment for the commercialization of creative ideas, and opening up more opportunities for exchanges and interplay among creative talent. The Government will continue to allocate resources to foster a rich variety of cultural and arts activities. ... We are prepared to consider helping them [the cultural sector and community organizations] set up a cultural and creative think tank to gather and groom more talent and experts in cultural and creative studies, who can work with the Government to promote the development of cultural and creative industries (Tsang 2005).

The following paragraph is found in his 2007 policy address:

I hope that the recently announced plan for the WKCD project will stimulate the development of cultural and creative industries in Hong Kong. At the same time, we need a large pool of creative talents and a discerning audience. To achieve this, the Education Bureau will encourage the nurturing of creativity, talents and ability in artistic and cultural appreciation in primary and secondary students, and promote university training for creative, performing arts and cultural talents (Tsang 2007).

As indicated in the above paragraph, the nurturing of creativity and ability in artistic and cultural appreciation among primary and secondary students was to be fostered for the sake of creating an audience pool for WKCD and the cultural and creative industries, which suggests that education could rotate around these spindles. Both Tung and Tsang considered culture and arts from their functional values, with Tung emphasizing the value of bridging East and West and Tsang emphasizing economic benefits.

In 2009 cultural and creative industries was identified as one of six emerging industries crucial to the development of Hong Kong's economy. The six emerging industries complement the four pillar industries: financial services, trading and logistics, tourism, and professional and producer services. The other five emerging industries are medical services, educational services, innovation and technology, testing and certification services, and environmental industries.

Tsang also mentioned a number of concrete projects, items, or plans in his policy addresses. For example, in 2006 he mentioned implementation of the Venue Partnership Scheme, in which operators of performance venues of the LCSD partnered with performing arts groups, and the founding of the Hong Kong Film Development Council. In 2009 he proposed the launching of Create Hong Kong and the CreateSmart Initiative. In 2011 he designated 2012 as Hong Kong Design Year and established the Bruce Lee Gallery in the Hong Kong Heritage Museum. These concrete items were rarely mentioned in Tung's cultural discourse. Filling his cultural content with concrete items reflects Tsang's administrative rationale, as he has long been an administrative officer in British colonial government.

In the paragraphs discussing cultural and creative industries in his policy addresses, Donald Tsang mentioned CEPA several times, with the film industry as a major topic. In 2009, he stated:

> Since the implementation of CEPA, preferential treatment has been given to our local film industry. ... Because of this advantage, we have seen notable growth and breakthroughs in the output, scale and box office receipts of Hong Kong–Mainland co-productions in the past few years. Of the top 10 box office hits in the Mainland in 2008, six were Hong Kong–Mainland co-productions (Tsang, 2009).

Tsang's major focus was on economic benefits. If C. H. Tung had delivered a message of the same kind, he might have added, "The coproductions of Hong Kong and mainland filmmakers have facilitated the cultural exchanges of the two places."

Another point worth noting about Tsang is that he mentioned bringing cultural activities into local communities in 2008 and 2010:

> To bring more cultural activities into local communities, the Government will encourage cultural and performing arts groups to stage performances across the territory (Tsang 2008).

> To allow arts and culture to reach out to the community, we will display in our parks, open spaces and government

offices buildings visual art pieces created by budding artists, students or teams, and improve the image, facilities and services of our public museums (Tsang 2010).

In fact, when the Urban Council and Regional Urban Council were abolished in 2000, the HKSAR government promised to let district councils play a bigger role in urban and cultural matters, but it turns out that cultural resources have long been in the hands of the government. Of course, Donald Tsang's mentioning of bringing arts and culture into the community was not about the decentralization of cultural resources to different districts of Hong Kong. However, in the cultural policy platform of the manifesto for the CE election of 2012, C. Y. Leung, the third CE, mentioned "delegating authority to the 18 districts: We will encourage the district councils to play a more prominent and proactive role in promoting culture and arts." The following is a summary of the cultural discourse of C. Y. Leung in his first policy address.

C. Y. Leung's Era

First, Leung picked up Tung's viewpoint that Hong Kong is a place of interaction between Eastern and Western cultures. But differences can be observed in their discourse. The following is found in Leung's first policy address:

> Hong Kong culture is a unique fusion of Chinese and western influences. We will strive to promote our arts and cultural development and related industries with local color to enrich our lives and raise our city's profile in the Mainland and abroad. ... With its geographical location and historical background, Hong Kong has a unique position in Chinese cultural history. Through a sophisticated fusion of Chinese and Western influences, it has created a pluralistic culture of its own. It has also achieved great success in preserving Chinese cultural heritage, such as Cantonese Opera. We should continue to leverage our strengths and advantages to further our mission (Leung 2013).

Tung put more emphasis on the function of Hong Kong as a bridge between East and West, while Leung focused on the local features of Hong Kong and the inheritance of Chinese traditional culture. Leung did not say that Hong Kong people should learn more about the culture of the motherland as Tung did. Instead, Leung mentioned cultural and creative industries and the West Kowloon Cultural District. But unlike Tsang, Leung said that culture and arts had their own discursive space. Concrete projects were proposed in Leung's cultural discourse, with most of them, such as the Arts Capacity Development Funding Scheme and the CreateSmart Initiative, being planned during Donald Tsang's era.

In Leung's first policy address, culture and arts were linked with land and housing, as indicated in the following paragraph under the title "Land Supply": "We need land for housing development; we also need land for elderly homes, students' hostels and venues for hosting sports, religious, arts and cultural events. ... Hong Kong still has many tracts of undeveloped land and potential sites for reclamation" (Leung 2013).

In this policy address, C. Y. Leung did not mention the setting up of a cultural bureau, which was his pledge in the manifesto. For his other pledge of "delegating authority to the 18 districts," he mentioned the following: "We plan to provide an additional $20.8 million a year ... to enhance the work of DCs [district councils] in promoting arts and cultural activities at the district level" (Leung 2013).

This was a new element that had not been mentioned in the policy addresses of the two former CEs. It can be anticipated that in the future, there will be discussion on whether cultural resources should be centralized in the hands of the government or decentralized to the districts.

CONCLUDING OBSERVATIONS

The three chief executives have written a total of sixteen policy addresses revealing the similarities and differences between their cultural discourses. Tung frequently mentioned that Hong Kong people should learn more about the culture of the motherland and that the cultural traits of Hong Kong allowed it to act as a bridge between Eastern and Western cultures. The discourse illustrated his political intention to bring about more exchange and understanding between the people of Hong Kong and those of its motherland. With terms such as *arts, culture, West Kowloon Cultural District, industries,* and *CEPA,* Tung set the tone for his two successors. Donald Tsang's policy addresses were full of concrete projects. His content regarding culture and the arts mainly concerned the cultural and creative industries and the West Kowloon Cultural District, with the focus on practical issues. Both Tung and Tsang perceived culture as a means to achieve another goal: for Tung, cultural exchange between Hong Kong and China; for Tsang, economic development. The word *arts* had disappeared totally for two years and was reborn in discourse about cultural and creative industries and the West Kowloon Cultural District.

In other words, economic development related to these aspects facilitated the reappearance of and emphasis on the term *arts* in the discourse of HKSAR leaders. Tsang, in particular, focused on economic development with the launching and realization of initiatives such as CEPA, the film industry, Create Hong Kong, the CreateSmart Initiative, and Hong Kong Design Year. Meanwhile, analysis of Leung's cultural discourse was based on only one policy address—his first one, delivered in 2013. Hence the findings are not as comprehensive as for the other two CEs, who gave more addresses. In general, Leung reiterated the East–West cultural exchange viewpoint of Tung and continued some concrete projects put forward in Tsang's administration. His newly added element was the role of district councils, which could have an impact on the cultural life of Hong Kong people if well furnished. The argument of whether cultural resources should be centralized in the hands of the government or decentralized to the districts could be a hot discussion topic. Together with the West Kowloon Cultural District, cultural and creative industries, and the postponed Cultural Bureau, these topics are likely to be central points of discussion in Hong Kong's future cultural discourse.

Note: This chapter is reprinted with permission of Professor J. Scott Goble and Professor Tadahiko Imada. An earlier version of this chapter appeared in *Cultural Inclusiveness and Transparency of Purpose in Contemporary Music Education: The Influence of Cultural, Educational, and Media Policies* (University of British Columbia, 2012).

BIBLIOGRAPHY

Leung, C. Y. 2013. "Seek change, maintain stability, serve the people with pragmatism" (2013 policy address), http://www.policyaddress.gov.hk/2013/eng/ (accessed December 6, 2013).

Tsang, D. 2005. "Strong governance for the people" (2005–2006 policy address), http://www.policyaddress.gov.hk/05-06/eng/index.htm (accessed December 6, 2013).

———. 2007. "A new direction for Hong Kong" (2007–2008 policy address), http://www.policyaddress.gov.hk/07-08/eng/policy.html (accessed December 6, 2013).

———. 2008. "Embracing new challenges" (2008–2009 policy address), http://www.policyaddress.gov.hk/08-09/eng/policy.html (accessed December 6, 2013).

———. 2009. "Breaking new ground together" (2009–2010 policy address), http://www.policyaddress.gov.hk/09-10/eng/index.html (accessed December 6, 2013).

———. 2010. "Sharing prosperity for a caring society" (2010–2011 policy address), http://www.policyaddress.gov.hk/10-11/eng/index.html (accessed December 6, 2013).

Tung, C. H. 1997. "Building Hong Kong for a new era" (1997 policy address), http://www.policyaddress.gov.hk/pa97/english/patext.htm (accessed December 6, 2013).

———. 1999. "Quality people, quality home, positioning Hong Kong for the 21st century" (1999 policy address), http://www.policyaddress.gov.hk/pa99/english/speech.htm (accessed December 6, 2013).

———. 2000. "Serving the community, sharing common goals" (2000 policy address), http://www.policyaddress.gov.hk/pa00/pa00_e.htm (accessed December 6, 2013).

———. 2003. "Capitalizing on our advantages, revitalizing our economy" (2003 policy address), http://www.policyaddress.gov.hk/pa03/eng/policy.htm (accessed December 6, 2013).

———. 2004. "Seizing opportunities for development, promoting people-based governance" (2004 policy address), http://www.policyaddress.gov.hk/pa04/eng/index.htm (accessed December 6, 2013).

The original version was published in Chinese and is published here with the permission of Hong Kong Economic Journal Company Ltd. Ref: 2014JUN06007.

Chapter 7

Challenges of Museums in Hong Kong in Promoting Local Cultural Heritage: An Assessment of Hong Kong International Museum Day (2001–2010)

Wai Shing Lee

The International Council of Museums (ICOM) updated the definition of museums in ICOM statutes in 2007 during the twenty-first general conference as follows: "A museum is a non-profit, permanent institution in the service of society and its development, open to the public, which acquires, conserves, researches, communicates and exhibits the tangible and intangible heritage of humanity and its environment for the purposes of education, study and enjoyment."[1] Museums exist to provide the public with not only mere exhibitions of antiques but also the diversity of other services listed in the definition. The mission of museums has been redefined constantly to meet public perceptions/expectations based on continual interaction with host societies. Due to this public service expectation, the first International Museum Day (IMD) was established in 1977 as a platform to link societies and museums to promote the importance of museums.

Hong Kong has been organizing IMD since 2001 to urge more participation in the public museum sector. As the ICOM proposed, preparation for IMD "provides the opportunity to draw the public's attention to museum programmes through activities taking place on and around 18th May."[2] Certainly, the museum and IMD are closely interrelated, with the museum's goal being to introduce cultural messages to the public while IMD is a tool to achieve that goal, so we may simply view them as two faces of the same coin. On the other hand, scholars such as Kenji Yoshida argue that the museum is not only a place to store and exhibit tangible cultural heritage but also a place to create intangible cultural heritage and to transmit it to the public.[3] Examining how to keep China's intangible heritage alive in museums, Pan Shouyong suggests various methods such as the introduction of digital and visual exhibitions.[4] Acknowledging the relationship between the museum and IMD and the connection between the museum and cultural heritage, this paper attempts to discuss impacts of Hong Kong International Museum Day (HKIMD) on Hong Kong's cultural heritage preservation and promotion.

MUSEUM EDUCATION AND THE DEVELOPMENT OF HKIMD

Enhancing public attention to cultural heritage to a certain extent means educating the public on identifying and appreciating heritage, which involves the issue of museum education. Museum education itself, as George Hein recognizes, is an old topic and has been a profession at least since World War II.[5] Through various exhibitions in museums, people can learn more about subjects such as the history of a particular period or impacts of a profound artist or philosopher. In general, government constructs a platform for the public to learn about its culture, society, and even identity. In this context, HKIMD was suggested in response to an alteration of the Hong Kong government's cultural policy.

In 2001 the newly founded Leisure and Cultural Services Department (LCSD) organized the first HKIMD, and this action was highly strategic because of a departmental reconstruction. Previously, the Urban Council and the Regional Urban Council, which had been independent from other Hong Kong government departments, administered cultural affairs, creating a separation of cultural policy and the government's central policy.[6] The cancellation of the two councils allowed the Hong Kong government to execute cultural affairs with collaboration with other departments. The rearrangement induced a policy shift to take more account of the audience and public. The LCSD, taking over all recreational and cultural affairs from the two former councils, aimed at a more aggressive approach to encourage public participation in museums and cultural heritage.

The LCSD employed a strategy of active cultural policy to get closer to the general public after its establishment. The two canceled councils handed over administration of museum, performing arts, and library services to the Cultural Services Branch under the newly established LCSD, hinting at a transition of administration as well as policy changes in these areas. In a meeting of the Legislative Council Panel on Home Affairs on December 14, 2001, the Home Affairs Bureau stated that the LCSD had founded two offices, the Audience Building Office and the Arts Promotion Office, in 2000 and 2001, respectively. The former was responsible for the promotion of interest in performing arts among the general public while the latter was in charge of visual arts. The setting of these two offices reflected the desire of LCSD to start a brand-new plan, different from that of the former urban councils, to actively and dynamically promote cultural affairs in Hong Kong society.

On September 1, 2001, the LCSD launched the School Culture Day Pilot Scheme, attempting to cultivate the general public from childhood. This was an example of the LCSD's new active strategy. The Home Affairs Bureau introduced the aim and contents of the scheme to the Legislative Council as follows:

> 30 schools have been selected and their students are being offered cultural and arts programmes especially designed for them in LCSD museums, performance venues and public libraries. The aim is to broaden their vision, developing their interest in arts and fostering their creativity.[7]

Schools joining the scheme could give students in-depth experiences through activities such as visiting LCSD performing venues, museums, and libraries during school hours. This scheme has been carried out by the Audience Building Office to promote culture and arts, described under the category "Arts Education Programmes" in the LCSD annual report.[8] The scheme was a manifestation of the strategy to get closer to the general public and to educate them rather than merely providing cultural services. But the efficiency of the scheme was relatively low compared with HKIMD because it targeted only the schools, while HKIMD targeted the whole society.In regard to the needs for arts education, LCSD organized HKIMD to develop knowledge of and interest in culture among the general public and also as a promotional media to construct an atmosphere favoring cultural events. In fact, the realization of the needs can be traced back to 1996, when the two urban councils still operated. A subcommittee of the Urban Council named the Culture Select Committee created a five-year plan, targeted on "expand[ing] the audience base for the arts" and "carry[ing] out audience education activities to promote the public's appreciation of the arts."[9] In 1998 the Hong Kong Arts Development Council, which since 1995 had served a supporting role for arts development in Hong Kong by planning programs and allocating funds, accepted directions for advancing and enhancing the content and quality of cultural education in schools and the community to provide the public with more information about culture and arts.[10] Even before the establishment of LCSD, the consensus about educating audiences in how to appreciate cultural performances and heritage had already been conceived.The LCSD continued the aim of arts education, and this was the motivation of organizing the first HKIMD. As stated by former secretary for home affairs Leo Kwan Wing-wah in a Legislative Council meeting in 2000, the plan of HKIMD clearly indicated that the Hong Kong government follow directions determined in the pre-LCSD period. He said:

All LCSD museums will join hands in organizing a promotion programme from 18 to 20 May 2001 to celebrate the International Museums' Day (18 May) in order to promote more visits to the museums and encourage the wider use of museum services. We will continue to enhance our museums' role in promoting "life-wide learning," including the establishment of resource centres in all major LCSD museums, a Children's Discovery Gallery in the new Hong Kong Heritage Museum, and so on. This will provide our school children and the public with more opportunities to understand and appreciate our art, history and cultural tradition.[11]According to Kwan's speech, the primary objective of HKIMD, however, was to encourage visits to LCSD museums. Enhancing participation in museum services seemed to be a lower priority. Emphasized by Kwan as the focus of museum services after the start of HKIMD, the slogan "life-wide learning" has tended not to be popular among—or at least not familiar to—the general public years later, which I will explain below, although the government regarded HKIMD as a means to educate the public on how to appreciate cultural heritage through museums. Hence HKIMD unintentionally served as a gimmick under which public awareness of the importance of museums was not a major concern. After several years, the original vision for HKIMD had slowly changed in terms of a demand for quantity of visits rather than quality of participation.Undoubtedly, the popularity of HKIMD in the museum sector is getting greater. According to statistics shown below, the number of units participating in HKIMD increased from 2001 to 2010.

Table 7.1 Numbers of participating units (2001–2010)

Year	Number of Units	Year	Number of Units
2001	22	2006	31
2002	24	2007	35
2003	26	2008	37
2004	28	2009	40
2005	30	2010	37

Sources: Hong Kong Special Administrative Region government press releases, May 11, 2001; May 6, 2002; May 10, 2003; May 15, 2004; April 29, 2005; April 27, 2006; April 20, 2007; April 28, 2008; May 16, 2009; May 1, 2010.

Twenty-two units joined the first HKIMD and the number rose to a peak in 2009, when forty units participated—almost double the number in the first year. The major reason for the increase was the involvement of private and non-LCSD museums and cultural institutions in HKIMD. During these ten years, only four museums operated by LCSD were founded.[12] At the same time, as shown in Table 7.2, there was a great increase in non-LCSD units participating, from six in the first year to seventeen in 2010—almost half the total number that particular year. Their participation definitely supported HKIMD. In other words, the growing popularity of HKIMD was based on private and non-LCSD museum participation, as the number of LCSD units kept constant at sixteen most of the time and later increased a little to twenty. Initially, HKIMD was for the promotion of LCSD museums. Thus the increase in non-LCSD units significantly implied creation of a cultural atmosphere by private units.

Table 7.2 Numbers of non-LCSD units (2001–2010)

Year	Number of Non-LCSD Units	Year	Number of Non-LCSD Units
2001	6	2006	14
2002	8	2007	17
2003	10	2008	18
2004	12	2009	21
2005	14	2010	17

Sources: Hong Kong Special Administrative Region government, press releases, May 11, 2001; May 6, 2002; May 10, 2003; May 15, 2004; April 29, 2005; April 27, 2006; April 20, 2007; April 28, 2008; May 16, 2009; May 1, 2010.

Having many private and non-LCSD units join together was unexpected, since the original aim of HKIMD was to encourage visits to LCSD museums. The involvement of these units was especially supportive for the event as well as the whole museum sector, because these institutions, including other governmental departments and private organizations, could furnish information that LCSD was not able to offer. For instance, LCSD had not suitably dealt with development of the Hong Kong police or the history of horse racing in Hong Kong because of resource restrictions by respective organizations. Therefore, museums from other departments and the private sector were necessary to provide the public with more comprehensive museum services, and HKIMD then played a role in exhibiting a variety of services in Hong Kong society. From this point of view, HKIMD served not only as a tool for LCSD museums to encourage visits but also as a platform for all other cultural institutions to promote themselves. Open in 2008, the Tao Heung Museum of Food Culture, for example, required advanced reservations for both individual and group visitors, which certainly was a barrier for people, inevitably weakening their intention to visit. As a privately owned museum, the Tao Heung Museum had limited resources to operate and propagandize, so could it make use of HKIMD to convey itself to and attract the general public, which did not know about it before or had not had an opportunity to visit due to the awkward formality. In another case, the Tung Wah Museum introduced the history of the Tung Wah Group of Hospitals and itself to the public at its booth at HKIMD 2008, with an emphasis on the number of participants visiting the booth.[13] The promotion of museums greatly underlies participation in HKIMD. Alternatively, non-LCSD institutions might make use of HKIMD to gain popularity and eventually to boost numbers of visitors rather than to concentrate on quality of museum content. LCSD museums, non-LCSD governmental department museums, and private institutions all had their own reasons to partake in HKIMD, but this annual event contributed to the integration of a variety of museum services for the general public in Hong Kong. Whether to educate the public or to boost the number of visits, museums engaged in HKIMD with activities and performances can easily attract the public during the event each year. From 2001 to 2008, the focal activity in HKIMD was the Museum Panorama, which was mostly held at the piazza of the Hong Kong Cultural Center, although some activities were held at the Hong Kong Science Museum and the Hong Kong Heritage Museum. In the Panorama, booths exhibiting museum information through various games and other attractions accomplished a promotional effect. For instance, the Flagstaff House Museum of Tea Ware brewed Chinese tea for visitors in 2007; the Hong Kong Racing Museum brought two miniponies for picture-taking with visitors in 2008. However, in 2009, under the theme of "Museums and Tourism," chosen by the ICOM, the Museum Panorama was unexpectedly replaced by a fleet of parade lorries around the city, with which the LCSD directly and aggressively communicated with tourists. An official description of the event below reveals an eagerness by government to promote Hong Kong museum services to welcome local visitors as well as foreign tourists:

> Under the theme of "Museums On the Go," a fleet of parade lorries carrying performance teams including museum mascots and international cultural ambassadors participated in shows for the public at the opening ceremony and gala held today at the Cultural Centre Piazza. The parade lorries will visit major tourist attractions and shopping areas on Hong Kong Island, in Kowloon and the New Territories tomorrow (May 17) from 10am to 6pm to show the rich offerings of the museums.[14]

Frontal communication with the general public became more frequent through a series of activities on HKIMD, but the contents of HKIMD did not relevantly fulfill the educational purpose set forth at the beginning. In every annual report of LCSD, the number of visitors was obviously emphasized to show a great use of museum services. Although it did not mean that LCSD was blindly seeking out achievement in the "hard" statistics, its listing of numbers slightly revealed a different vision of cultivating public knowledge and appreciation of arts. As seen in the examples above, the enrichment of museum education through HKIMD was not apparently seen in each item; rather, promotions of museums to attract more visitors were easily seen. The nature of IMD in Hong Kong has been altered. The success of HKIMD did not equal a flourishing museum sector in Hong Kong because the number of attendees does not simply reflect the quality of cultural enrichment, let alone an increasing awareness of cultural heritage.

SUCCESS OR FAILURE? INFLUENCE OF HKIMD

HKIMD had a direct influence on tourism instead of cultural heritage itself. HKIMD enriched the tourist industry with more attractions apart from consumption in shopping malls. In fact, the museum sector is turning out to be a part of cultural heritage tourism. As Bruce Prideaux and Dallen J. Timothy pointed out, although culture and heritage may be ordinary and familiar to local people, they may be unique and special to foreign visitors, and cultural heritage has become a significant selling point of tourism.[15] In Hong Kong, museums could serve as spots of cultural heritage tourism. A newspaper clipping in 2004 showed that foreign visitors enjoyed HKIMD and remarked that it was the right direction for the development of Hong Kong tourism:

> It was the first time for Mr and Mrs Zon from Canada to travel Hong Kong and they felt extremely excited to the arts and culture of Hong Kong. In the "Museum Panorama" yesterday, they visited the booth of Hong Kong Heritage Museum. ...They were interest in the plentiful contents of Panorama and thought Hong Kong have to organize more similar activities, or even something like "cultural travel" to introduce the characteristics of arts and culture in Hong Kong.[16]

A Guangzhou citizen expressed a similar view: "The consumption habit of visiting Hong Kong was shopping in the past. But now the living standard in Mainland has been enormously improved, so I now go to Hong Kong not only for shopping but also for feeling the cultural ambiance."[17] The above comments reveal the possibility of diversifying Hong Kong tourism by adding cultural travel, in which museums as well as HKIMD play a role in promoting tourism. In 2004 Guangzhou organized IMD as well, and Hong Kong sent six museum representatives to advertise Hong Kong cultural and historical resources for Guangzhou citizens. Tom Ming, former chief curator of the Hong Kong Heritage Museum, said that the museum wished to attract more mainland tourists to Hong Kong cultural institutions, including museums, through this kind of promotion.[18] The relationship between IMD and tourism was very clear, although this was an example from Guangzhou, also reflecting a cultural exchange among mainland China, Hong Kong, and Macao. Apart from the influence on tourism, HKIMD was an occasion for Hong Kong museums to exchange with overseas institutions, especially on the mainland and Macao. As mentioned, Hong Kong has joined the Guangzhou IMD. HKIMD has also welcomed participation by foreign institutions. Since 2003 the Macau Museum has participated in HKIMD; the Guangdong Provincial Department of Culture, the Beijing Municipal Administration of Cultural Heritage, and other organizations have joined in since. This participation positively deepens a mutual understanding of cultural services and resources, opening the possibility for cultural institutions to organize mega-events. A report submitted by the Committee of Museums in Hong Kong in 2007 said that establishment of partnerships with overseas museums could benefit organizations with world-class exhibitions, which could help promote the image of Hong Kong museums.[19] Indeed, external communication with foreign cultural institutions has been a major business and a chief issue of LCSD since the first LCSD annual report in 2000. The exhibition *The Pride of China: Masterpieces of Chinese Painting and Calligraphy of the Jin, Tang and Yuan Dynasties*, held in 2007, was an example of collaboration with a museum from the mainland. This event attracted attention to one of the most renowned antiques, *Along the River during the Qingming Festival*, and encouraged the general public to learn about paintings in ancient China. Hence HKIMD was a platform for exchange and dialogue between Hong Kong museums and foreign cultural organizations, consequently widening the horizons of the public in Hong Kong. However, awareness of cultural heritage itself remains low.To examine how HKIMD has stimulated awareness of cultural heritage, I first discuss the effect of HKIMD on the museum sector. It might be valid to say that HKIMD is a commodity and that the general public is a consumer as a metaphor for the public's perception of museums in Hong Kong. As former secretary for home affairs Kwan mentioned in his speech, one of the aims of HKIMD was to enhance the role of "life-wide learning" via LCSD museums, providing the public with educational resources. Therefore, how HKIMD helped spread knowledge of arts and culture in society is a standard by which to review whether it has succeeded in the above aim. In November 2004, LCSD appointed Mercado Solutions Association Limited to conduct a survey covering seven museums. Both frequent and infrequent visitors were surveyed in an effort to understand their characteristics as well as factors that encouraged them to visit museums.[20] Although the survey was conducted in 2004, when HKIMD was just four years old, it still indicated the effectiveness of HKIMD. Applying an exit-interview approach for frequent visitors, the survey successfully interviewed 5,021 visitors during the first two weeks of November, with about 718 cases at each museum on average. The awareness level and participation rate in educational and public programs at each museum varied, as shown in Figure 7.1.

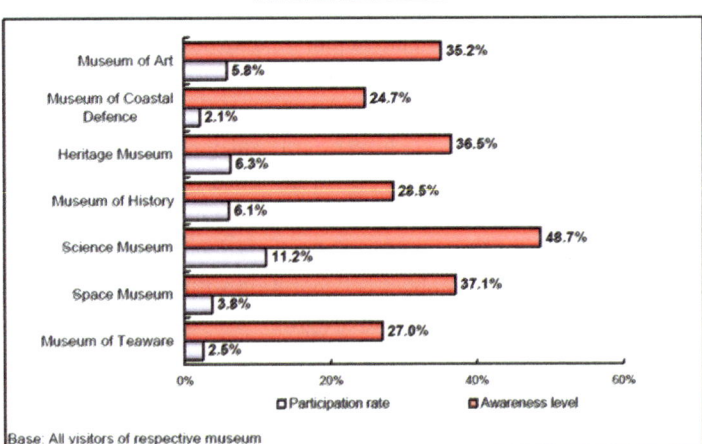

Source: Leisure and Cultural Services Department. "Visitors' Survey on Major Museums under Leisure and Cultural Services Department." Prepared for the Museums Advisory Group, Consultative Committee on the Core Arts and Cultural Facilities of the West Kowloon Cultural District, p. 13. LC WKCD-321.

Figure 7.1 Awareness level and participation rate of education and public programmes organized by the museums

The Science Museum acquired 48.7 percent of the visitors, realizing its educational and public programming goals, while other museums had lower numbers, ranging from 24.7 to 37.1 percent. Participation rates were much lower, with the highest of 11.2 percent for the Science Museum and the lowest at 2.1 percent for the Hong Kong Museum of Coastal Defense. Based on these percentages, the situation of museum services in Hong Kong was fairly poor, reflecting the fact that the educational role of museums was not widely known, despite the objective of HKIMD to promote "life-wide learning" in society. It was obviously a contrast to the Hong Kong government's motivation to organize HKIMD. At least among major LCSD museums, more than half of frequent visitors did not understand any arrangement of educational activities, and less than 10 percent took part in activities. More importantly, even for frequent visitors, the survey results were relatively poor. For infrequent visitors, the results were even more unsatisfactory. Another purpose of HKIMD was to boost the use of museum services, but the result was the opposite. The same group randomly interviewed infrequent visitors—those who had not visited any of the seven museums in the previous four years—

by telephone, with 1,004 successful interviews of 3,300 contacted individuals. The results showed that around one-third of the interviewees had not visited one of the seven museums in the prior four years. In spite of the four-time HKIMD, around 30 percent of the people had not visited any of the major LCSD museums. If the infrequent visitors had already participated in HKIMD, this may imply that HKIMD had not achieved its goals in increasing visits to museums. It may even be considered a failure of HKIMD that this annual event could not bring the public to use more museum services than before. Moreover, 21.7 percent of the 1,004 infrequent visitors knew about free admission to LCSD museums every Wednesday, and 5.6 percent of them understood that annual passes, half-year passes, and weekly passes were for sale. More importantly, only 0.6 percent of infrequent visitors had participated in the Museum Panorama, and this totally defied efforts of HKIMD to help people enjoy museum services.[21] The data signify a difficulty in attracting people who were not interested in visiting museums. Therefore, the objectives of HKIMD to enhance museum services were seriously questioned. The influence of HKIMD seemed to be much lower among infrequent visitors. To ascertain why infrequent visitors had not utilized museum services, they were asked in telephone interviews about factors affecting their decision to visit museums. The survey conductor provided them with eight possible choices and let them answer yes or no in each category. The final result is reported in Figure 7.2.

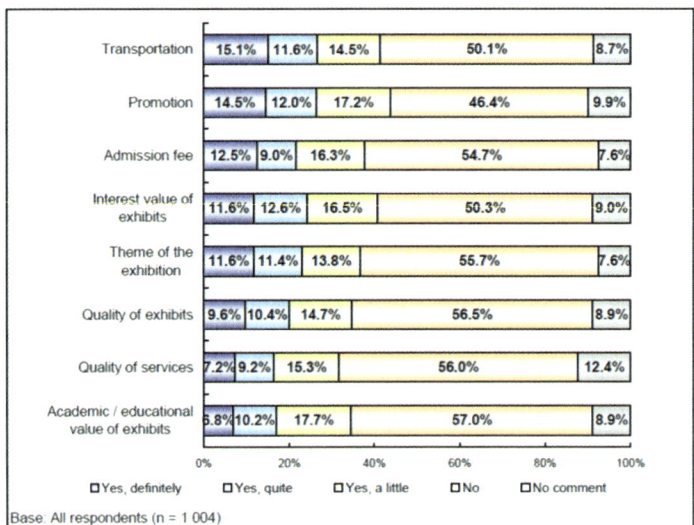

Source: Leisure and Cultural Services Department. "Visitors' Survey on Major Museums under Leisure and Cultural Services Department." Prepared for the Museums Advisory Group, Consultative Committee on the Core Arts and Cultural Facilities of the West Kowloon Cultural District, p. 21. LC No. WKCD-321.

Figure 7.2 Factors affecting the decision of visit to the museums

For every possible factor leading infrequent visitors not to visit museums, around half of the interviewees answered no—ranging from 56.3 to 68.4 percent if "no comment" was also included. It means the choices provide real alternatives for them but the survey results reflect that all choices were not directly related to why they had not visited museums. This was definitely obscure. Also, it is worth noting that half of the infrequent visitors (50.5 percent) thought the advertising of museums was very or quite inadequate; 15.6 percent thought the opposite, and 16.2 percent considered the advertising to be average.[22] HKIMD is the largest annual event of LCSD to promote museum services, but parts of the general public perceived nothing and even expressed a negative impression of museum services, regardless of how keenly LCSD pushed the use of its museum through HKIMD. The report, which might be a relevant reference for planning museum development in the future, indicated several shortcomings of museum services, such as the low participation rate and low public awareness level. The evidence in the report almost rejected the efforts of LCSD, which attempted to encourage wider use of museums through HKIMD, because it could not effectively influence most of the general public, no matter if they were frequent or infrequent visitors. Since the direct effect of HKIMD was largely on tourism and the public perception of museums had not changed much due to the organization of HKIMD, it was a question of how this largest annual event in the museum sector had benefited cultural heritage in Hong Kong.

BENEFITS TO CULTURAL HERITAGE

IMD, as defined by the ICOM, it is supposed to draw the public's attention to museum programs and the role the museum plays in preserving the tangible and intangible heritage of humanity and in educating society to inherit cultural assets. In other words, IMD is held for the inheritance of culture. However, it was a challenge to revive cultural heritage in Hong Kong through HKIMD.In spite of the increasing prosperity of HKIMD, it did not denote a continuous and accumulative enhancement of the public's interest in museums, especially in cultural heritage. Fluctuations in numbers of visitors to seven museums from 2002 to 2010 (Table 7.3) signify a limitation of Hong Kong museums in the inheritance of heritage. As shown in the table, both the Museum of Art and the Museum of History recorded a sudden increase in numbers of visitors in 2007 and 2003, respectively, because of conspicuous exhibitions: The Pride of China: Masterpieces of Chinese Painting and Calligraphy of the Jin, Tang, Song and Yuan Dynasties from the Palace Museum and Treasures of the World's Cultures from the British Museum at the Museum of Art in 2007; and *War and Peace: Treasures of the Qin* and *Han Dynasties and Napoleon Bonaparte: Emperor and Man* at the Museum of History in 2003. Borrowing materials from famous overseas museums to hold the exhibitions, the two museums highlighted special topics, which easily fascinated audiences because the materials were distinguished worldwide and were seldom loaned overseas. After these mega-events, numbers of visitors dropped, which could be attributed to public interest in prominent international collections only. It was incorrect that holding a world-class exhibition would be much more efficient than organizing HKIMD for the promotion of cultural heritage, so the relationship between HKIMD and the revival of heritage was uncertain.

Table 7.3 Numbers of visitors to seven museums (2002–2011)

Year	Art	Coastal Defense	Heritage	History	Science	Space	Tea Ware
2002	298,000	331,650	745,000	897,000	868,000	750,800	N/A
2003	217,000	213,895	455,629	1,017,240	870,000	655,800	N/A
2004	270,000	182,300	632,000	658,000	850,000	787,800	206,000
2005	547,000	177,533	657,000	562,750	967,000	765,482	195,000
2006	410,000	161,896	510,000	581,329	797,975	703,572	212,800
2007–08	621,682	152,024	474,393	618,976	1,184,335	781,981	182,241
2008–09	294,727	130,925	393,535	688,743	1,045,431	803,524	167,000
2009–10	338,067	125,940	478,578	627,138	1,003,227	678,785	163,460
2010–11	358,290	122,883	397,575	614,032	1,838,577	423,146	169,913

Source: Leisure and Cultural Services Department, annual reports 2002–2010.

Taking the Hong Kong Museum of Heritage as an example, the revival of cultural heritage through HKIMD was not easily proved. Displaying antiques such as items used by renowned opera artists in an exhibition named *Cantonese Opera Heritage Hall,* the museum claimed to offer insight into the history and unique characteristics of Cantonese opera. The museum set up a corresponding booth in HKIMD, letting people try on opera costumes to experience the unique cultural assets of Hong Kong and Guangdong. However, it is doubtful whether these activities and programs have really cultivated the public regarding opera and created an atmosphere of participation. According to the Home Affairs Bureau, Cantonese opera was included in the core as well as the elective module of music curriculum under the New Senior Secondary Curriculum of 2009.[23] In the first Hong Kong Diploma of Secondary Education (HKDSE), among 448 students who originally chose to study music in the new curriculum, 240 eventually went for the exam, with 71,800 total students sitting for the first HKDSE.[24] This meant that around 0.6 percent of candidates approached the opera, and only around 0.3 percent sat for the public exam. If teenagers could not easily learn traditional opera in the formal curriculum, it would likely be very difficult for them to learn from extracurricular activities, including HKIMD. In other words, one of the aims of HKIMD was to help promote heritage in society, but the effect was not clearly reflected at the school level because, in the case of the Cantonese opera, the corresponding atmosphere of engaging in cultural heritage was not established accordingly. This showed the difference between the reality of and the desire for educating or enriching the public about the appreciation of arts through HKIMD. Although the HKDSE could not represent the entire inheritance of Cantonese opera, it indicated the problem of low participation rates in arts and cultural heritage as well. On the other hand, popular response to activities like the opera costumes in the Museum Panorama during HKIMD seemed contradictory to the above notion of unconcern for cultural heritage, possibly due to the speedy disappearance of enthusiasm for cultural heritage among the general public after HKIMD because of different attitudes toward HKIMD and museums by the public and the government. After ten years of HKIMD, public attitudes toward this annual event have gradually diverged from the Hong Kong government's desire to boost the use of quality museum services. The main reason for this divergence was different understandings of the concept of leisure, in which scholars have already discussed and conceptualized two categories: serious and casual. Robert Stebbins defines serious leisure to be constituted of amateurism, hobbyist activities, and volunteering, by which individuals reward themselves both personally and socially in ten dimensions, such as personal enrichment, self-actualization, social attraction, and group accomplishment. Casual leisure is an immediate, intrinsically rewarding, relatively short-lived pleasurable activity requiring little or no special training to enjoy—for example, play, relaxation, and active entertainment.[25] Casual leisure, as Tony Blackshaw points out, "is associated with shallowness and superficiality and serious leisure is associated with depth and substance."[26] These definitions of leisure could also be employed in explaining different understandings of HKIMD as well as cultural heritage. Parts of the general public that always actively participate in cultural activities or museums programs could be viewed as exercising serious leisure because they enjoy benefits such as gaining cherished experiences (personal enrichment) or associating with other participants to enlarge their social worlds (social attraction). Back to the case of the Cantonese opera, more and more teenagers and even children are becoming interested in this traditional opera. They are motivated to develop an outstanding skill for a better academic future or to release stress via opera. Contrarily, those who try on opera costumes during HKIMD but have seldom or never partaken of any related activities may well consider this cultural heritage as play or entertainment for the weekend, which is a short-lived and immediate pleasure. They most likely would not take part in the opera later. Indeed, this was quite common in most programs in HKIMD, symbolizing HKIMD as casual leisure. A personal experience from a professor in Hong Kong, Kai-Wing Chow, who has been concerned about the development of museums in Hong Kong, showed that HKIMD tended to be casual leisure for most people. Chow participated in HKIMD 2008, describing the Museum Panorama as "moderately crowded but the throng of people did not evenly distribute to every booth, some having a long queue while some others only two or three visitors or even being desolate."[27] It was quite clear that some institutions were relatively welcome and popular compared with others. Chow also discovered that "outside the booth of the Hong Kong Museum of Medical Sciences was very crowded because people could get a free blood measure and the elderly were the majority." He noted differences in sizes of crowds, as visitors could get special benefits from joining particular units. It was also not surprising that booths with games were relatively attractive to audiences. HKIMD, at least the Museum Panorama, was seen as entertainment for most of the general public—as a chance to relax. Other activities, like visiting museums for free, were casual leisure for most people as well, with Figure 7.1 giving strong evidence that regular visits to museums were not popular. Because of the divergence of attitudes toward HKIMD between the public and the Hong Kong government—the former treating it as a casual activity with the latter regarding it as a means to deepen cultural knowledge—HKIMD could not be expected to transmit cultural heritage if the public saw the event as casual instead of serious. The revival of cultural heritage could not benefit merely from museum services as well as HKIMD. The main reason was different understandings about activities provided by museums. Lots of people in Hong Kong viewed HKIMD as casual leisure during the weekend rather than as a chance to academically enrich their cultural knowledge. Referring to Figure 7.2, the almost total unconcern for museum services was apparently visible, as nothing besides those eight comprehensive choices could suitably explain why infrequent visitors did not visit museums. For infrequent visitors, visiting museums is casual leisure and a short-term pleasure, so they did not visit museums when they had other casual activities to replace museum services. It was probably a dilemma for HKIMD to identify itself as casual or serious leisure. If it was serious leisure, it would certainly face the same results as museum services as shown in Figure 7.2; if it was casual leisure, it could attract people for immediate pleasure but not for knowledge enrichment. Therefore, so far cultural heritage in Hong Kong cannot be easily revived or promoted by HKIMD, which is, to a great extent, casual leisure in public perception.

CONCLUDING OBSERVATIONS

HKIMD could not effectively raise attention to and interest in cultural heritage among the public in Hong Kong. As IMD was originally organized for drawing the public's attention to museum programs, one should judge the efficiency of HKIMD based on feedback to the programs and eventually cultural heritage from the general public rather than on the prosperity of activities in one or two days. In spite of being purposeful and constructive, HKIMD has not contributed much to the promotion or revival of cultural heritage, but it did influence the general public in perceiving museum services as casual leisure.Although the aim of HKIMD was to boost museum usage, participation in museum programs remained low—unchanged when compared to the time before the first HKIMD. Due to the predominance of casual leisure and the public preference for mega-exhibitions, HKIMD has not greatly impacted regular museum visits, a contradiction to the purpose of HKIMD. Following its ineffectiveness in encouraging use of museum services, HKIMD certainly could not influence cultural inheritance. The majority of participants in HKIMD treated the Museum Panorama as entertainment, just like a carnival where they could play games provided by different institutions. But they did not seriously care about the purpose of HKIMD. Notwithstanding the divergent attitudes toward HKIMD, cultural development in Hong Kong has not been discontinued.Cultural development in Hong Kong in these ten years has advanced, partly because of HKIMD. Aiming at assessing the relationship between HKIMD and the revival of cultural heritage in society, this paper demonstrates the ineffectiveness of HKIMD on the one hand. On the other hand, even though HKIMD has been treated as casual leisure, it did raise attention to the term *cultural heritage* in the past decade. HKIMD made little contribution to cultural heritage in Hong Kong, but at least it stimulated discussions about the preservation and conservation of cultural heritage in society. In short, HKIMD during the study period of this paper has had less impact on cultural heritage promotion and preservation, which is unsatisfactory in terms of the aims of IMD, but in the case of Hong Kong, HKIMD has done little to prompt an understanding of cultural heritage.

APPENDIX

Museums Participating in Hong Kong International Museum Day (2001–2010)

Museums under the Leisure and Culture Services Department
1. Antiquities and Monuments Office (2001)*
2. Art Promotion Office (2001)
3. Central Conservation Section (2001)
4. Dr. Sun Yatsen Museum (2007)
5. Fireboat Alexander Grantham Exhibition Gallery (2009)
6. Flagstaff House Museum of Tea Ware (2001)
7. Hong Kong Film Archive (2001)
8. Hong Kong Heritage Discovery Center (2006)
9. Hong Kong Heritage Museum (2001)
10. Hong Kong Museum of Art (2001)
11. Hong Kong Museum of Coastal Defense (2001)
12. Hong Kong Museum of History (2001)
13. Hong Kong Railway Museum (2001)
14. Hong Kong Science Museum (2001)
15. Hong Kong Space Museum (2001)
16. Law Uk Folk Museum (2001)
17. Lei Cheng Uk Han Tomb Museum (2001)
18. Ping Shan Tang Clan Gallery cum Heritage Trail Visitors Center (2008)
19. Sam Tung Uk Museum (2001)
20. Sheung Yiu Folk Museum (2001)

Museums of Other Hong Kong Governmental Departments
1. Hong Kong Correctional Services Museum (2004)
2. Hong Kong Planning and Infrastructure Exhibition Gallery (2004)
3. Hong Kong Wetland Park (2007)
4. Lions Nature Education Center (2003)
5. Police Museum (2001)
6. Hong Kong Jockey Club Drug InfoCentre (2010)

Privately Owned Museums
1. Art Museum of the Chinese University of Hong Kong (2001)
2. Dr. and Mrs. Hung Hin Shiu Museum of Chinese Medicine of Hong Kong Baptist University (2008)

3. Foods of Mankind Museum (2005)**
4. Hong Kong Maritime Museum (2006)
5. Hong Kong Museum of Medical Sciences (2001)
6. Museum of Education, Hong Kong Institute of Education (2009)
7. Museum of Ethnology (2001)
8. Po Leung Kuk Museum (2002)
9. Stephen Hui Geological Museum of the University of Hong Kong (2009)
10. Tao Heung Museum of Food Culture (2009)
11. Hong Kong Racing Museum (2001)
12. Tung Wah Museum (2002)
13. University Museum and Art Gallery of the University of Hong Kong (2001)

Other Museums and Institutions
1. Beijing Municipal Administration of Cultural Heritage (2007)
2. Cultural Affairs Bureau of the Macao Special Administrative Region Government (2006)
3. Guangdong Provincial Department of Culture (2004)
4. Guangzhou Municipal Bureau of Culture (2005)
5. Macau Museum (2003)

*   Numbers in parentheses indicate the joining year for each museum.
** It operated from 2002 to 2005. Tao Heung Holdings Limited continued running it as Tao Heung Foods of Mankind Museum from October 2005. In 2008 it moved to Fo Tan as the Tao Heung Museum of Food Culture.

NOTES

1. "Museum Definition," International Council of Museums, http://icom.museum/who-we-are/the-vision/museum-definition.html (accessed July 18, 2012).
2. International Council of Museums, "International Museum Day," *ICOM News* 54, no. 1 (2001): 4, http://icom.museum/fileadmin/user_upload/pdf/imd/2001/2001-1_eng.pdf (accessed July 18, 2012).
3. Kenji Yoshida, "The Museum and the Intangible Cultural Heritage," *Museum International* 56 (2004): 110.
4. Pan Shouyong, "Museums and the Protection of Cultural Intangible Heritage," *Museum International* 60 (2008): 17–18.
5. George E. Hein, "Museum Education," in *The Companion of Museum Studies*, ed. Sharon Macdonald (West Sussex, UK: Wiley-Blackwell, 2011), 340–52.
6. For more information of the councils and their cancelation, see Runhe Liu, *A History of the Municipal Councils of Hong Kong: 1883—1999: From the Sanitary Board to the Urban Council and the Regional Urban Council* (Hong Kong: Leisure and Cultural Services Department, 2002).
7. Home Affairs Bureau, "Review of the Supernumerary Deputy Director (Administration) and Assistant Director (Finance) Posts in the Leisure and Cultural Services Department," enclosure 2, page 3. Meeting of the Legislative Council Panel of Home Affairs on December 14, 2001. LC paper CB(2)667/01-02(03). Legislative Council, Hong Kong.
8. Leisure and Cultural Service Department, *Annual Report 2002*, http://www.lcsd.gov.hk/dept/annual2002/en/cultural02.php (accessed July 18, 2012).
9. Urban Council's Culture Select Committee, *The Five-Year Plan of the Urban Council's Culture Select Committee Consultation Paper* (Hong Kong: Urban Council, 1996), 12.
10. Hong Kong Arts Development Council, *The Study Report and Integration of Comprehensive Abstracts of the Definition, Implementation and Resources Development of Hong Kong Cultural and Arts Policy* (Hong Kong: Hong Kong Arts Development Council, 1999), 104–105.
11. "Legislative Council Official Record of Proceedings on 22 November 2000" (Hong Kong: Legislative Council, 2000), 1137–38.

12. They were the Hong Kong Heritage Discovery Center (opened in 2005), the Dr. Sun Yat-sen Museum (2006), the Ping Shan Tang Clan Gallery cum Heritage Trail Visitors Center (2007), and the Fireboat Alexander Grantham Exhibition Gallery (2007).
13. Tung Wah Group of Hospitals, "Events of the Tung Wah Museum," *Tung Wah News*, August 2008, 6.
14. Hong Kong Special Administrative Region Government, "International Museum Day 2009 Opens Today," press release, May 16, 2009.
15. Bruce Prideaux and Dallen J. Timothy, "Themes in Cultural and Heritage Tourism in Asia and the Pacific," in *Cultural and Heritage Tourism in Asia and the Pacific*, ed. Bruce Prideaux, Dallen J. Timothy, and Kaye Chon (Oxon, UK: Routledge, 2008), 1–14.
16. "Foreigners felt interested, they like Hong Kong culture and customs, visiting museums like joining a cultural travel," *WenWeiPo*, May 19, 2001, A14.
17. "Promote Cultural Travel, Cultural Institution in Hong Kong and Macao Going to Guangzhou for Promotion," *Ta Kung Pao*, May 18, 2004, B02.
18. Ibid.
19. Committee of Museums, "Recommendation Report," meeting of the Legislative Council Panel on Home Affairs on June 8, 2007, p. 31. LC paper CB(2)2042/06-07(05). Legislative Council, Hong Kong.
20. The seven museums included the Hong Kong Museum of Arts, Hong Kong Museum of History, Hong Kong Science Museum, Hong Kong Space Museum, Hong Kong Heritage Museum, Hong Kong Museum of Coastal Defense, and Flagstaff House Museum of Tea Ware.
21. Leisure and Cultural Services Department, "Visitors' Survey on Major Museums under Leisure and Cultural Services Department," information for the Museums Advisory Group, Consultative Committee on the Core Arts and Cultural Facilities of the West Kowloon Cultural District, p. 20. LC paper WKCD-321. Legislative Council, Hong Kong.
22. Leisure and Cultural Services Department, "Visitors' Survey on Major Museums under Leisure and Cultural Services Department," p. 22. LC paper WKCD-321. Legislative Council, Hong Kong.

23. Home Affairs Bureau, "The Promotion of Cantonese Opera and Other Xiqu," meeting of the Legislative Council Panel on Home Affairs on March 14, 2008, p. 6. LC paper CB(2)1310/07-08(03). Legislative Council, Hong Kong.
24. "8% students forgive sitting for the HKDSE," *Ming Pao*, December 2, 2011, http://life.mingpao.com/htm/hkdse/cfm/news3.cfm?File=20111202/news/gfa1.txt (assessed May 21, 2012).
25. Lee Davidson and Robert A. Stebbins, *Serious Leisure and Nature: Sustainable Consumption in the Outdoors* (Basingstoke: Palgrave Macmillan, 2011), 8–17.
26. Tony Blackshaw, *Leisure* (London: Routledge, 2010), 41.
27. Kai-Wing Chow, "A Record of HKIMD2008," *Contemporary Historical Review* 9, no. 2 (2008): 33–35.

Chapter 8

Revival of Chinese Calligraphy in Hong Kong's Architecture

Roger Tin Sing Kho

Following the handover of sovereignty of Hong Kong to China, there has been a constant engagement in identity searching. This awakening regarding the historical role of Hong Kong has nourished a resurgence in Chinese art and culture. And the fermentation of culture, whether driven by polity or the brick-and-mortar accumulation of people, is the steady infusion of different cultures that creates mutual assimilation and vibrant mingling of ideals.

During colonial rule, lots of Chinese traditions were still being practiced by local Chinese and new migrants from mainland China. The strong resilience against foreign acculturation could be seen in all aspects of life. Social activities in the form of traditional rituals, ceremonies, festivals, and funerals were just a few examples. Calligraphy in different formats was used to convey a variety of symbolic ideals. The spring couplet (春聯) was used to make auspicious wishing; the column couplet (楹聯) to guard entrances against evil spirits; the horizontal tablet (牌匾) to acknowledge inauguration of business; epitaphs (碑銘) to mark the foundation of buildings; and even temporary bamboo floral decoration (花牌) to celebrate festive occasions. Ideals were thus embodied in calligraphy as literary content, style, script type, choice of material, and other historical referents or hearsay. Chinese calligraphers have left an abundance of works in different styles within different edifices, public and private. In these works of calligraphy, calligraphers not only presented self-identity through their personal styles to commemorate special occasions but also presented stylistic and historical references to the long tradition of calligraphy dating back to early calligraphers in different dynasties.

This emphasis on genealogical lineage of famous master calligraphers is a particular practice in Chinese calligraphy. The symbolic content of a work of calligraphy is a complicated network woven by a series of historical anecdotes and personal predilections. With such an extensive boundary and flexibility for creativity, Chinese culture was able to assimilate foreign ideals into its melting pot and survive different periods of political uncertainties.

With the advent of the electronic age, Chinese calligraphy is facing another serious challenge—the demise of handwriting. Yet in recent years, there has been a surge in the number of people practicing Chinese calligraphy both in China and Hong Kong. What caused this revival is still unclear, but growth in economic power must have played a vital role in the development of soft power in culture. With the help of technology and the economy, calligraphic relics and archaeological findings were uncovered. Images on calligraphic works of ancient masters were retouched and revitalized in colorful publications. This development has increased the popularity of Chinese calligraphy as an art form both in China and around the world. And this revival in Chinese calligraphy has also generated international attention by a number of Chinese scholars.[1] Dialogues were created with a global vision to investigate the identity of Chinese calligraphy from different angles overseas and within Greater China.[2] Such revitalization has improved the recognition of Chinese calligraphy as a cultural element. And a discourse on treating calligraphy as writing and as art would be of significance in a globalized setting. To appreciate calligraphy as graphic patterns of art without an awareness of the textual content and underlying symbolism would weaken the diversified identity of the art. To counterbalance the proliferation of electronic devices and the subsequent neglect of handwriting, a return to the value of traditional practice in Chinese calligraphy would definitely improve cultural development.

Calligraphy is an inscription of man in space. A study into this art form in the context of a city would bring back the essence of art to the people in their environment. Using Kevin Lynch's idea about the legibility of city space and the special situation in the local city context, a richer network of mental maps of calligraphic works can be built. Lynch divides city elements into paths, edges, districts, nodes, and landmarks.[3] Paths are channels that allow people to travel in the city, such as roads, trails, and tram lines. Edges are perceived boundaries such as coastlines and city walls. Districts are relatively large areas of the city distinguished by some identity or characteristic, such as the Ngong Ping area featuring a number of Buddhist sites. Nodes are focal points, intersections, or loci. Landmarks are physical objects that serve as external reference points such as the stone gateway in Sun Yatsen Memorial Park. Through the distinct features of this city, people create mental maps of what the city contains. And the mental representation, along with its physical presence, creates identity for a place and its people. This innovative approach in understanding the way people read calligraphy works in city spaces reveals new ways of identity-building and place-making.

Chinese characters started as pictograms. Thus the use of symbols in Chinese culture can be seen in all aspects of life. Symbolic flourishes are applied in literature, architecture, painting, and calligraphy in China. And the practice creates a network of intertextuality that requires a high level of erudition to decipher. This sharing of common identity through symbolism and its daily application in culture are important contemporary issues. The study into calligraphy as symbol in the social context of architecture creates a richer background for discussion. And the implication for a globalized China as a culture hub is of considerable interest.

Importantly, putting calligraphy work back into the context of architecture encourages art appreciation. It also improves creative sites for calligrapher-artists through a richer context for art exhibition in comparison to the current displaying of discrete objects in museums. Presently, there is a trend to display art objects out of museum constraints in an open-trail experience. Chinese calligraphy, with its long history and traditions, has showed its resilience in different eras. It has been able to assimilate different ideals and can be seen flourishing in the city space of Hong Kong.

HISTORICAL LANDMARK AND PATH CREATION

After the 1997 return of Hong Kong to China, there were heated debates in the political scene on a variety of issues. A simplified dichotomy of pro-China or anti-China was often used, yet the content of what contributed to the Chinese identity was generally overlooked. Under such circumstances, all relevant traces of a commonly accepted historical figure, Sun Yatsen, and his activities in Hong Kong were collected and displayed to rebuild a comprehensive local history. Sun Yatsen was the first president and the founding father of China. He is usually aptly referred to as the father of the nation. Thus the symbolic meanings provided by reproductions or representations of his writing, ideals, physical appearance, history, and even calligraphy and handwriting would all initiate a significant referent to the history of modern China. Due to the fact that most of the landmark buildings related to history have been demolished over time, a re-creation of landmarks was of utmost importance.

Sun Yatsen Historical Trail started in November 1996 under a local consultative body within the Central and Western District Board. The trail has fifteen spots, with the starting point at Hong Kong University and the finishing point at the original site of He Ji Zhan (和記棧) at D'Aguilar Street. In 2001 the Hong Kong SAR government decided to devote major efforts to promote various historical monuments in Hong Kong and allocated funds for the provision of new plaques for the Sun Yatsen Historical Trail. Two additional spots were added to the original thirteen spots when Kom Tong Hall (甘棠第), used to house the Dr. Sun Yatsen Museum, was formally opened in 2006.

Pak Tze Lane Park (百子里公園, 2011), as the eighth stop on the Dr. Sun Yatsen Historical Trail, was home to the Furen Literary Society (輔仁文社), founded in 1892. In 1895 the Furen Literary Society merged with Dr Sun's Revive China Society (興中會, Xinzhonghui), with help from Yau Lit (尤列). Yang Quyun (楊衢雲) became president and Sun Yatsen secretary of the Revive China Society. Due to its strategic location, Pak Tze Lane was a frequent meeting place for revolutionaries; many plots against the Qing imperial government were conceived here.

Pak Tze Lane Park was opened in 2011 to commemorate the centennial anniversary of the 1911 Chinese Revolution. All the literary content, sculptural figures about the cutting of pigtails, and exhibition panels were unified recollections on the theme of the origin of the Chinese Revolution. The traditional theme of Chinese gardens was revamped into this subtle re-creation of history as a cultural symbol. The pavilion, with new architectural elements made of recycled timber slips, was used to present the ideals of the Furen Literary Society. The most dramatic presentation was conceived in a semicursive calligraphy inscription written by Sun Yatsen in a mourning letter on the assassination of Yang Quyun[4]

Figure 8.1

To amplify the emotional content of the mourning over Yang Quyun's death, the semicursive script runs in a comparatively unrestrained manner on a series of recycled timber slips. The letter was originally meant to be read in hand but was enlarged to a monumental scale. This transformation to a large format on a series of recycled timber slips in a small park has drastically increased the visual impact of the original handwriting in a congested urban location flanked by buildings and alleyways on four sides. The traces of emotion in Sun Yatsen's brush strokes can be felt strongly in the representation, but the structural layout of his calligraphy was radically distorted with embossed characters in the form of a bas-relief. The symbolic meaning of Pak Tze Lane of early years in the history of modern China was thus re-created with the inclusion of this calligraphy of modern representation.

Another monumental construction, Sun Yatsen Memorial Park (中山紀念公園 2010), was originally named Western Park (西區公園). In 2002 the government renamed it. The location of the park has no particular significant value in any historical incidents related to the 1911 Chinese Revolution. But because of its location near the harbor front and being close to the Sun Yatsen Historical Trail, a reconstruction and renaming of the park would effectively strengthen the symbolic meaning in terms of its urban connotation. At that time, in 2002, a giant rock with red calligraphy displaying the political ideal "The world is for all" (天下為公, tian xia wei gong) from The Book of Rites was installed on the west end of the park

Figure 8.2

The park was partially closed from early 2008 for construction of the Sun Yatsen Memorial Park and Swimming Pool Complex, which was completed in June 2010. Additional facilities included a Chinese stone gateway in the park entrance near Connaught Road West on the south

Figure 8.3

The four Chinese characters of the political ideal "The world is for all" were used again as the dominant motif. And the horizontal tablet in the middle of this three-tier stone gateway was strategically located in line with a bronze effigy of Sun on an open lawn. With the map of the Central and Western District carved on a polished granite floor, the mental map for the locations of activities of Sun Yatsen was thus formally visualized on a grand scale.

In this project, the Chinese calligraphy of Sun Yatsen plays an important symbolic role in the creation of the language of this urban landmark. The standard script used can be easily recognized because of the square and orderly form, as opposed to the more spontaneous cursive script adopted in Pak Tze Lane Park. With its simplicity and uprightness, the standard script surely facilitates projecting Sun Yatsen's political ideals to a larger audience. The four large characters in Chinese calligraphy, appropriated by Sun from ancient classics into his political ideal, have a very strong connotation for Chinese identity-building. And the repetitive use in different locations creates a strong mental representation in city space, which is further reinforced by the openness of an urban park.

Compared to the modern reinterpretation in Pak Tze Lane Park, Sun Yatsen Memorial Park uses the more traditional architectural style of Pai Lau (牌樓), an urban landmark that has a stronger symbolic than functional meaning. And the calligraphy on the giant rock and on Pai Lau shows a slight difference from the brush stroke used by Sun Yatsen. The characters on the giant rock give an impression of square brush strokes, while those on the new addition possess more rounded brush strokes. As pointed out by Qing politician and calligrapher Kang Youwei, "Lifting results in elegance and coherence, pressing results in preciseness. Rounded brushstroke is dispersed and free; square brushstroke is congealed, orderly and calm."[5] The forceful and upright brush strokes are then associated with his determination in the formulation of his democratic ideals.

URBAN PARKLANDS OF HISTORY

By putting into discussion the history of two urban parks constructed in different periods—one in the 1980s and the other in the 1990s during colonial rule—we can see how Chinese historical elements were incorporated into the park design. Both parks were designed as Chinese gardens, but the history behind each site is significantly different. By looking into the historical facts uncovered from the site and how Chinese elements—particularly Chinese calligraphy—were presented, we can see how calligraphy was appropriated as a symbol in different circumstances.

Hollywood Road Park (荷里活道公園, circa late 1980s) was originally marked as Possession Point (水坑口) on a map in the 1980s and before. This was where the British army landed on Hong Kong in 1841 before the signing of the Treaty of Nanking. It was formally marked as Possession Point with a raising of the Union Jack and a gun ceremony. Later, in the 1960s, it became a popular bazaar known as Tai Tat Tei (大笪地, large piece of land). It was nicknamed the civilian's nightclub (平民夜總會) and housed a number of old Hong Kong–style street stalls and entertainment activities. With traditional entertainment such a Cantonese opera, storytelling, fortune telling, and kung fu performance, the area was vibrant, with people enjoying cheap food at night.

In the name of urban renewal, Hollywood Road Park was designed as a Chinese garden to improve the sanitary and noise situation around the area. Red columns, green glazed tiles, a pavilion, a pool, stone features, and white a stucco wall with latticed tracery—all elements of strong

Chinese symbolism—were used to cover up the earlier history. And the red-columned gateway with Chinese calligraphy in seal script on stone

Figure 8.4

was completed with the renaming of Possession Point as Hollywood Road Peak, a name with a neutral connotation. The only reference to the site's history was a display panel of old photos. Because of the less legible seal script, a horizontal stone tablet with standard script inscription was erected on the entrance in a less prominent location.

Figure 8.5

The natural and archaic simplicity of the calligraphy in standard script can be easily traced back to the calligraphy of Zhong You (鍾繇).[6]

Zhong You was a calligrapher and politician in the state of Wei during the Three Kingdoms period. He was adept at clerical, semicursive, and standard script and was known as the father of standard script. The typical style of his calligraphy can be seen in Memorial on an Announcement to Sun Chuan (宣示表, Xuanshì biao) and Memorial Recommending Ji Zhi (薦季直表, Jian Ji Zhi biao). The script style evokes a big contrast to the colonial history of the site. And the meaning of this urban park in its new physical form is completed with archaic symbolism in Chinese calligraphy.

Another urban park, the Kowloon Walled City Park (九龍寨城公園), designed in Chinese garden style, was constructed in the mid-1990s, the last few years of colonial rule. The Kowloon Walled City was notorious for its illegal structures of high density and a virtual lack of civil order. It was demolished in April 1994 before the handover of Hong Kong. The agreement between the colonial government and Chinese authorities was to clear the area and build a park on the site. The idea was to incorporate the remaining building features of historical value into a Jiangnan garden of the early Qing Dynasty style. The original wall had been demolished during the Japanese occupation, and a rewarding number of relics were unearthed during park construction and were subsequently included in the design. As a result, the history was carefully preserved, with great attention to strengthening the underlying historical symbolism.

Two carved granite plaques unearthed from the original South Gate, bearing calligraphy in standard Chinese characters for "South Gate" (南門) and "Kowloon Walled City" (九龍寨城), were carefully preserved as remnants strongly symbolizing the ravaging of Chinese authority

Figure 8.6

This symbolism of an earlier military presence was amplified with the fist calligraphy (拳書) of the characters shou (壽).

Figure 8.7

and moyuan (墨緣)

Figure 8.8

by Zhang Yutang (張玉堂), commodore of the Dapeng Brigade, carved on a stone tablet. Together with other relics recovered from the site, including three old cannons, stone lintels, couplets, and column bases, a unified picture of Chinese sovereignty in the old days was then completed.

In addition to the rich content of history, new stone epitaphs of the narrative on the Kowloon Walled City Park were added both at the South Gate and the North Gate. Both epitaphs were written in standard script to record the history of reconstruction of the park, by Chen Shuheng (陳樹衡) on the shadow wall at the South Gate and by the architect Tse Shunkai (謝順佳) at the North Gate.

Figure 8.9

A number of other calligraphers also contributed to the collection of Chinese calligraphy in various script types and formats. Thus even with only the physical landmark of the imperial office of Yemen, the site history of the congested lawless enclave and festering squatter slum was replaced with earlier symbols of a military post re-created in a park.

Comparing the two projects by the colonial government, one can easily identify a considerable improvement in the use of calligraphy in the representation of site history. Hollywood Road Park, with its notorious history of British invasion, was intentionally downplayed to an urban green land with a Chinese overtone. Rather than an eclectic choice from an ancient master to create a new paradigm, the calligraphy used in the Kowloon Walled City Park was more carefully organized to reinforce the underlying classical concept and symbol of military outpost.

DISTRICTS OF RELIGIOUS WORSHIP

The essence of Chinese culture can be found in the religions of Confucianism, Taoism, and Buddhism. The three religions have a long history in China, and they share common symbols. Many of the relics of Chinese calligraphy are manuscripts related to these religions. Temples constructed to house the Buddha, historical figures deified, and legendary gods and goddesses were exuberant with interwoven ideals from the three religions.

The Wisdom Path (心經簡林) was conceived in 2002 when Jao Tsung-I (饒宗頤) completed the Heart Sutra calligraphy and dedicated it to the people of Hong Kong. The calligraphy work was much inspired by his visit to the Buddhist stone carvings on Mount Taishan in Shandong in 1980. The 260 Chinese characters in clerical script, with each character measuring six hundred by six hundred millimeters, were presented on thirty-eight wooden columns of eight to ten meters in height.

Figure 8.10

The columns were arranged according to the topography of the landscape in the shape of the infinity sign. The outdoor large-scale wooden inscription display resembles the bamboo slips used for writing in the Han Dynasty. As the original inspiration was from stone carvings on Mount Taishan, the calligraphy style shows what Jao had absorbed from the ancient stone carvings.

Figure 8.11

The style of brush stroke used is different from that on a Han bamboo slip and possesses a greater similarity to works in another Buddhist temple.

A similar calligraphy style can be found in the Ten Thousand Buddhas Monastery (萬佛寺,1957), with the Chinese characters "Ten Thousand Buddhas" (萬佛) written by Huang Wei Chang (黃維琯). The calligraphy work was presented as a giant horizontal tablet, using the entire facade of the temple hall as a sign. The calligraphy displays a rounded brush stroke that is dispersed and free. The style can be readily linked back to the ancient tradition in line with the Taishan jingshiyu (泰山經石峪).[7] This close relationship between the style in calligraphy and its religious ideal is readily found in many examples of calligraphy in temples. Taishan jingshiyu is a quintessential example of a rounded brush stroke, which is closely related to the Buddhist tradition.

In the Wisdom Path, the calligraphy of the Heart Sutra was written in large character writing (榜書, bangshu) to increase legibility. According to Kang Youwei, large character writing is significantly different from small character writing in the wielding and movement of the brush, body position, perfection in copying, and brush tip manipulation.[8] Due to these constraints, the presentation and techniques of small character writing in bamboo slip are drastically different from those of large character writing in stone carving. This arbitrary association of the format of presentation with the calligraphy style may be due to considerations of other technical issues regarding site selection, material constraints, and design considerations.

Hau Wong Temple in Kowloon City (九龍城侯王廟; refurbished and reopened in 2006) is located at the northwestern corner of Kowloon Walled City Park. The temple was built around 1730 to commemorate Sung general Hau Wong (Marquis Prince) Yeung Leung Jit (楊亮節), who protected the last emperor of the Southern Sung Dynasty from Yuan troops.The temple houses a wealth of cultural relics such as a stone epitaph carved in 1822, a wooden horizontal tablet and stone carving from the Qing Dynasty, and brightly colored Shek Wan pottery. During the Japanese occupation of 1941 to 1945, the rock around the temple area was quarried to extend the nearby Kai Tak Airfield. As a large amount of granite was needed, the stone wall of the Kowloon Walled City was torn down in the same period. This was the occasion when a stone carving of the Chinese "goose" character on the rock face was destroyed.

In 2005 the Chinese Temple Committee renovated the temple and reconstructed some of its lost relics. A series of Chinese calligraphy by contemporary calligraphers was added in a Poetry Garden and the destroyed stone carving was reinstated as a granite stone pavilion. The existing "goose," in one stroke (一字鵝), is a reconstruction of the former rock carving.

Figure 8.12

It is now presented on a flat wall in a triptych arrangement with couplets flanking the central character. The "crane," in one stroke (一字鶴) with a Taoist connotation, is sited behind the main hall of the temple. The stone carvings of "goose" in one stroke and "crane" in one stroke were made by Zhang Shou Ren (張壽仁) and Feng Shan (鳳山), respectively. Calligraphy is used here as a religious symbol in the form of a ropelike character to signify the power to tie up evil spirits. And the use of red as a substitution for peach wood refers to its magical power to drive away bad luck. The apotropaic property of a one-stroke character to ward off evil can be dated to the Han Dynasty.[9] The couplets flanking the "goose" in one stroke, written by Li Qing (黎慶), read:

古石書鵝摹逸少，
名山駕鶴仰侯王。
Imitate the calligraphy of "goose" by Yishao (Wang Xizhi) on the

archaic stone, Revere the loyal Hau Wong (Marquis Prince)
while riding a crane
over the famous mountain.

The literary content readily refers to the sage of calligraphy (書聖) Wang Xizhi (王羲之), using his alias Yishao (逸少), and the story about his hobby of rearing geese.[10] A legend says that he learned the key to turning his wrist while writing by observing how geese moved their necks. As a result of the parallelism in poetry, the geographical location, historical figures, calligraphy, and Taoist ideals are all brought together in one place to create a unified identity.

As the practice of three religions in China has adopted a pluralistic approach, interplay of religious ideals can be seen in many temples. This sharing of common symbols or practices is seen in calligraphy works displayed in the temple and religious site, and this cross-referencing has enriched the ground for identity-building.

NODES OF COMMERCE

In the early days of colonial rule, the influx of Chinese migrants due to political unrest in China brought calligraphers with a strong commitment to traditional Chinese culture to Hong Kong. To support themselves in a foreign land, they had to run different business or use their skills in calligraphy to earn a living. Under such circumstances, some calligraphers engaged in writing calligraphy for signage. Relics of their writings can be readily found in various locations in the city.

Calligraphy associated with private business generally originated from a simple need to signify the inauguration of business or a continuation of traditions. Calligraphy was usually affiliated with people from the same ancestral origin or with a close group of migrants to evoke certain emotions regarding the memory of their homeland. Examples were found in businesses such as teahouses, antiques shop, Chinese dried seafood shops (海味), and clan associations

Figure 8.13

(groups of migrants with a strong bond or a common identity). Calligraphy was used to display common ideals and blessings. The display was always located in a prominent location to be unveiled on an auspicious day to strengthen the mental visualization. Yee Hing Loong (義興隆) was inaugurated in the trading of rare classic furniture and Chinese arts and crafts in 1945.

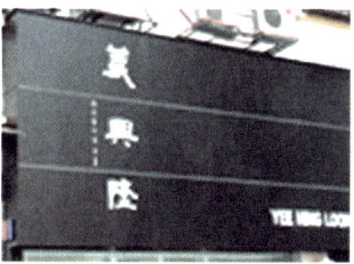

Figure 8.14

The calligrapher Luo Shuzhong (羅叔重) was a native of Nanhai (南海) in Guangdong Province who moved to Hong Kong in 1923. His clerical script originated from Northern Dynasties beike (inscriptions on stone tablets). With an aura of "metal and stone," it shows his attachment to the ideals of the stele school of the Qing Dynasty. Regarding Chinese calligraphy as shop signage, the most prolific calligrapher was Ou Jiangku (區建公), a native of Xinhui (新會) in Guangdong Province. In addition to his role as a calligrapher, he was also a practitioner of Chinese medicine. His calligraphy shows much influence from Qing Dynasty calligrapher Zhao Zhiqian (趙之謙) and his study of stele inscriptions from the Northern Wei, Han, and other early periods. He has produced extensive calligraphic shop signs for a noodle shop, a bookstore, a Chinese medicine company, a metal and machine tool company, and even a coffin shop.

Figure 8.15

The heavy strokes and triangular endings are easily recognized as inspired by Northern Wei stele inscriptions. Looking into these works of Chinese calligraphy in public spaces, one realizes the widespread application of calligraphy in daily Chinese life. In such small projects, the calligraphers had freedom to indulge personal expressiveness in their calligraphy ideals. And they treasured every way of exhibiting their works, even on building facades.

Figure 8.16

The style chosen by the calligrapher was often the external symbol of an affiliation with ancient calligraphy masters. In their pursuit of perfection in calligraphy, they allowed their brush strokes to flow freely in the cityscape, waiting to be identified.

CALLIGRAPHY AS A SIGN IN THE CITY

The use of Chinese calligraphy in architecture has undergone a significant revival since the return of Hong Kong to China. Looking into calligraphy in Hong Kong as a city element reveals how images are linked to create a legible series of historical reconstruction. The creation of these social connotations is an important key to the understanding of identity-building in a place. The earliest study on signs can be dated to the works of Ferdinand de Saussure and Gestalt visual psychology. Later works by Kevin Lynch on the image of cities and Robert Venturi's book on symbolism in architecture have influenced the architectural field. In the Han Dynasty, the rules for construction of Chinese characters were categorized into six principles of character formation, or "Six Writings" (六書, liushu). A comprehensive consideration of these origins of symbolic creation has enhanced a deeper understanding on the subject.

The pictographic content of Chinese characters is in itself a sign language. The formation of Chinese characters, their origins, the use and transformation in script style during different dynasties, and the formats of presentation were all consequences of the ideology and power structures in society. The pictograms（象形字）, ideograms（指事字）, and phono-semantic compounds（形聲字）, with intricate differences, are the basic units to creating networks of intertextual reference. From works of calligraphy, we have seen political ideals linked to Confucian classics, historical figures related to each other with common identities, religious belief transformed into a style of brush stroke, color and material usage empowered with magical function, and personal style exemplified as an adorable model. All these delicate variations in the final presentation of calligraphy are the key to deciphering the embedded symbols. Trained calligraphers are often very conscious of the embedded ideals in the movement of their brushes, which is the outcome of lifelong dedication. With a thorough study into the mechanism used to create identity, Chinese calligraphy can be used as an effective medium to convey the message of a civilization.

NOTES

1. Qiu Zhenzhong, ed., *Calligraphy and Chinese Society* (Beijing: Beijing Normal University Press, 2008).
2. Wang Yuechuan, *The Identity of Calligraphy* (Beijing: Peking University Press, 2008).
3. Kevin Lynch, *The Image of the City* (Cambridge, MA: MIT Press, 1960).
4. For the use of letters as samples of Chinese calligraphy, see Qianshen Bai, "Chinese Letters: Private Words Made Public" in *The Embodied Image: Chinese Calligraphy from the John B. Elliott Collection*, ed. Robert E. Harris and Wen C. Fong (New York: Harry N. Abrams, 1999), 381–99.
5. Kang Youwei, *Guang yizhou shuangji* (Taipei: Jinfeng, 1999), 286.
6. Zhong You (鍾繇) and Zhang Zhi (張芝), the Han-era master of the cursive script, along with two Wangs (Wang Xizhi, 王羲之, and his son Wang Xianzhi, 王獻之), have been honored by latter-day scholars as the "Four Worthies" of the calligraphic world.
7. Taishan jingshiyu jingangjing belongs to the genre of Moya Keshi (摩崖刻石) and was carved on the riverb**ank** about one kilometer northeast of Doumugong (斗母宫) in Taishan, Shandong. It is one of the largest Moya Keshi Buddhist sutras in China. The texts were carved on a gentle sloping stone platform in an area of 2,064 square meters. There are forty-four columns in total, with each column having 10 to 125 characters, totaling 2,799 characters, with character size around fifty centimeters.
8. Kang Youwei, *Guang yizhou shuangji*, 318–22.
9. Regarding "one-stroke" (i-bi-shu), see Cary Y. Liu, "Calligraphic Couplets as Manifestations of Deities and Markers of Buildings" in *The Embodied Image: Chinese Calligraphy from the John B. Elliott Collection*, ed. Robert E. Harris and Wen C. Fong (New York: Harry N. Abrams, 1999), 363–64.
10. Geese readily connect with the calligraphic style of Wang Xizhi, as Mi Fu during his forties wrote a poem stating that he had "decided not to play with geese." Cao Baolin, "Self-Expression versus Ancient Tradition: Calligraphy of the Song, Jin, and Yuan Dynasties," in Ouyang Zhongshi et al., *Chinese Calligraphy*, trans. and ed. Wang Youfen (New Haven: Yale University Press; Beijing: Foreign Languages Press, 2008), 261.

Chapter 9

Venturing Strategies in Developing the Hong Kong Art Industry

Yan Tung

Strategic partnership has been investigated as one of the most effective ways of sustaining a cultural industry as described in most studies (Cashman, 2010; Grams and Farrell, 2008; Rensburg, 2013; Saperstein and Rouach, 2002). It is linked with community practice, marketing, and philanthropy to help the organization connect to its target audience in creative programming. Successful strategies can accumulate "social and cultural capital," meaning capital that may not be money-related for organizations, such as good branding with a broad volunteer and audience base that helps attract funding and partnerships (Wagner, 2012).[1] Other promotional strategies such as utilization of social media tools reflect the communication pattern of an organization, from which we can identify the key communicators in a network (Mould and Joel, 2010; Pow et al., 2012). The above factors cradle the development of the production value chain in social and cultural entrepreneurship (Bilton and Cummings, 2010; Power and Allen, 2004). Innovation in mobilizing social and cultural capital can help the industry address community and society needs, which can generate profits in the long run to sustain industries in the art or nonprofit sector.

From the perspective of an art practitioner, this chapter will sum up a number of strategies employed by established cultural enterprises in Hong Kong for the past five years. This helps to formulate a structured analysis of the growth process of venturing strategies employed in the sector. My methodology is to investigate similarities in the established process that reflect common challenges encountered in Hong Kong. I base my observations on privately funded companies that have established their own foundations and on government-funded organizations. Five areas are found to be effective in these strategies: (1) geographical proximity/multicities operation strategies; (2) promotional strategies such as branding, signature events, and marketing; (3) networking strategies using media liaisons, stakeholders, and key management staff; (4) establishing foundations and school collaborations; and (5) investments in art to build up collections.

GEOGRAPHICAL PROXIMITY AND PLACE-MAKING

In Hong Kong, a small city with about 7 million people within 1,108 square kilometers of land,[2] place-making is a key factor in success.[3] To attract newcomers, it is vital that a place be well designed and suitable for showcasing art—particularly for places like old factories or barren construction sites that have not been associated with art. A place that is connected to public transport and neighborhoods will be frequently visited by the public when it is easily accessible. Bilton (2007, 2010) has described the success of creative clusters as a place-making practice to foster creativity. However, due to political and economic changes such as rental price fluctuation, there are only clustering effects for key flagship projects, which has marginalized other subcultures. In most art centers there will be ongoing projects, like providing artist housing, which can build up audience taste and capacity when there are more art activities. Preece (2011) proposed a similar idea on providing "locational advantage" to artistic people who would beautify their environment and place art where they have more opportunities to create. Increasing site accessibility will help them facilitate the entrepreneurial formation process.

The Central District on Hong Kong Island is the concentration hub for most international galleries. In the most expensive buildings, above the mass transit railway station and with high ceilings for contemporary artworks, the rent is affordable only for top galleries owned by patrons or collectors.[2] The art here is entirely commercial, with limited educational intent; audiences are target visitors who are potential buyers or traders. However, the sectors are connected, as artists travel across many platforms, and good new works are also available in nongallery contexts. It is common for sought-after artists to present works in different art organizations.

When organizations are located in a densely populated zone on the Kowloon side, they share similar audience composition. Locations near the mass transit railway can bring in a lot of tourists, including both target audiences and passersby who may not intentionally be looking for art. The influx of a large amount of traffic is a natural stage for performances and exhibitions that by nature need to develop audiences and can attract people who may not distinguish between decorative art and promotional activities. How the public perceives a place determines who visits the events, defining the identity of organizations and hence their potential to transform the environment to be more art tolerant.

When an organization is located in a mixed residential and industrial area, it will be visited mostly by working-class people. Visitors may find it exciting to discover art in industrial zones that are not normally associated with "high art," such as factory parking lots and roads full of trucks and trolleys. Other advantages include low rent for large spaces, with big cargo lifts for carrying, loading, and accommodating large artworks. There are also fewer complaints about noise from art and music events in industrial areas. These are all valid reasons for private and small art organizations to locate in factory areas. In proximity to artists' studios, factory sites can demystify the art-making process, which helps tempt the public to start engaging in art.

Some art organizations require multibillion-dollar investments over five to ten years. Surprisingly, both privately and publicly funded enterprises have built highly flexible long-term plans based on an uncertain political climate that can abruptly affect the direction of art projects. Smaller organizations have infrastructure budgets to fully cover setup costs but have to operate with either private sponsorship or project-based government funding. They do not have a stable programming focus or promotional budgets and have to create local markets to build up future audience bases. When the organization is privately owned, the planning is not only more dynamic and able to adjust to current market trends but is also free of government influence in the planning stage. For example, some have launched art marts and studios for local designers and artists in the past few years with a reasonably good response from the public. When an organization has to relocate, it can build other branches from client bases in several cities. Some have launched Chinese outlets focusing on media art and design; the network in China is branded like the one in Hong Kong. For publicly funded organizations, a good architectural plan designed by a star architect can brand the building as a landmark. The site can be well promoted to the public before the building is completed. In sum, there are many ways of sustaining a good branding image, provided that organizations can have flexibility in their business planning and distribution networks. When organizations can utilize the clustering effect and choose their locations, they can engage audiences in creative ways as mentioned previously in terms of geographical proximity.

PROMOTIONAL STRATEGIES: MEDIA CHANNELS AND RESPONSE FROM THE CIRCLE

Art media flourishes when there are demands for regular advertisements and advertorial needs by museums, galleries, art fairs, and other small to medium-sized exhibitors and performers. Before the flourishing times of art advertisements, there were only a few newspapers in the cultural field. Only recently have more English-language magazines and newspapers appeared in the market.[3] Some independent and Internet-based "underground" writing has initiated discussion groups on hot local issues, such as funding and governance structure, which resonates with student-run magazines.[4] Social media tools like Facebook are effective mechanisms by which to organize campaigns, as most artists are connected via these groups and can survey alternative opinions in closed chat rooms. In addition to local groups, overseas researchers and academics also visit websites on Hong Kong's development, and these informal platforms are transparent at all levels, even though records are not kept in official archives and often lack trustworthy sources.

A closer look at how art organizations utilize social media channels in promoting their activities reveals several phenomena. Large-scale organizations normally have a marketing department that constantly monitors Facebook activities and immediately responds to public inquiries via social media. By regularly posting events or "eye candy" such as images or lifestyle trend items, groups have successfully engaged a young and loyal audience who are IT literate and stay tuned when a promotional activity is going on. Both positive and negative comments lead to publicity. Maintaining an open and meaningful discussion on social media is determined by a public relations department, which at times has to handle crises, such as when the public opposes a politically sensitive campaign. Although social media normally runs on individual accounts, any comment posted under the company heading is no different from that on a regular media channel. Well-manipulated online platforms are essential to maintaining a good public image for organizations. Quite often they can help build a positive identity of the spokesperson in pronouncing semiofficial news.

For smaller organizations, media liaison is based on point-to-point and individually based contacts that cannot be sustained when staff turnover is high. Officers often share news regarding the organization, as they are in the same social media circle. It is time- and cost-effective to distribute news using social media platforms. However, people tend to be overreliant on social media, which causes confusion and overattendance when events are popular and widely shared. Small organizations can save advertising costs for print media by using free publicity tools like bloggers and can ask clients to post events albums, which is another effective way of sustaining client relationships. Art marketing becomes a trend in sustaining the momentum of a campaign and hence development of an organization. Technology, if used properly, is a powerful promotional tool for establishing good branding and building up an audience base, as well as substantially saving on advertising and printing costs.

NETWORKING

In social network analysis, certain key communicators serve to connect several industries in a sector. The functions of chief executive officer (CEO), as defined by Bilton and Cummings (2010: 160, 2010), are to set boundaries about how organizations should liaise with external stakeholders and prioritize their demands. This is termed a mapping and linking function. It is a power structure by which stakeholders can influence the strategies employed and the level of interest they have in supporting a particular strategy. Those who have a high level of power and great interest in influencing strategies are regarded as key players. The second function of the CEO is "sussing out" the values and capabilities of other people, formulating trust and teamwork, and encouraging them to promote the company.

Social media analysis can show how the CEO maps, links, and promotes the organization according to the different stages of development in a creative industry.[5] The first stage is recognition of ideas and events that can be evaluated and eventually developed into strategies before they are launched. It depicts the natural stages by which a business grows and declines via the characteristics of the leader. His personality combined with the right venture and industry experience can define the success of the business (Bilton and Cummings, 2010: 117). Preece (2011) identified three areas for a successful entrepreneur: a solid understanding of how the art form operates in the market, with business skills and a passion to overcome risk and uncertainty in the production value chain. The organization benefits from effective leadership over the long run.

This chapter explores two different social media, Facebook and LinkedIn, to unveil the complexities of the management structure of leadership and staff in the art industry. CEOs and their companies' websites show the following number of connections:

- Privately funded organizations with marketing departments: official page: 102 privately funded organzations, 34 connections

- Privately funded organizations with no marketing department: CEO: no connection: 1,378 (official page): 3,193 connections

- Publicly funded organizations with marketing departments: official page: 15,298; key management staff: 4,614

- Publicly funded organizations with no marketing department: CEO: no connections: 1,042 (official key event): 6,897 connections.

The figures show that organizations need key signature events and a marketing department to build up their audiences. Some rely on individual staff to build up key communication networks to disseminate news more efficiently. When an official page is managed by an assigned department, the figure is much higher than a page that relies on an individual staff member. The CEO is not necessarily the most networked person. Key staff members have more crossover networks. This shows the need for research into interconnected artistic leadership and collaborators in cultural entrepreneurship rather than singling out heroic individuals, and the need to look into explicit strategies (Bilton, 2010; Leung, 2014).

PARTNERSHIP

For commercial enterprises, under the current Hong Kong business registration ordinance, there are certain limitations to launching art activities such as a free exhibition open to the public.[6] Collaboration with schools or nonprofit organizations can cut down on costs and broaden an audience base, which is commonly used to build up the brand and public perception. Consequently, most art organizations have a nonprofit arm that designs outreach and educational programs to engage young people as volunteers or participants in large-scale events. These can be combined with internship programs that offer an entry point for newcomers and can link with schools that provide training in art disciplines and are able to produce content and periphery programs. These strategies are employed in all art organizations, some on a paid basis and others on an event basis.

When an organization sets up a foundation, it can take on more exchange activities with overseas institutions, more applications for government funding, and more art purchases.

As stated in Hausmann (2010), those starting up cultural businesses must think about business planning, patrons and sponsors, cash flow, distribution and media strategies, and a market review with audience analysis. Schools and university collaborations can bring more stable income to the company regardless of its commercial application, as is the case for all organizations. This can provide low labor costs and a team of knowledgeable volunteers to cover the lack of resources, pushing schedules and building up a loyal audience. Today, education is linked with the market for students, and skills provided by the market are vital to the survival of all art industries. Recent trends for artists to engage with community and educational projects have shown that art can be made accessible to the public without challenging its integrity.

FINANCE: PRIVATE AND PUBLIC; COLLECTION AS SOCIAL AND CULTURAL CAPITAL

Art is priceless because we collect intellectual output across different generations. Art theorists and historians cannot predict how artists will make their work. Archival knowledge is necessary when certain art movements or momentum is fading away. Institutions, like individual collectors, know the network and the market before they invest. It is usually their passion for the arts that prompts them to do real business. Collections are built out of individual taste and connections. It takes years to build up a valuable collection with political and market influence, especially if works are safely and properly kept in good condition. Artworks have high resale value in the market, for auction houses, museums, and other collectors. Many organizations house key collections worth billions. Collections can protect arts organizations when they lack funding, as artworks increase in value over time and can generate profits in auctions or other sales.

MULTI-CITIES ESTABLISHMENT AND LOCALIZATION

For organizations based in Hong Kong, planning begins when the flagship branch is in operation and other branches are mainly located in China or Southeast Asia. The business owners have already established networks and client bases in other cities, drawing on local resources to run their businesses effectively. Planning a multi-cities operation can give an impression to the public that the enterprise is well funded and widely networked to facilitate exchange in the long term. This is evident in many international galleries, shopping malls, and art education centers in Asia.

The multi-cities setup of publicly funded organizations relies on the recruitment of an international staff and consultants. Speaking the same language in the right culture is an important step to successful networking and collaboration. Those organizations that have a strong China network rely on consultancy and strong support from the company's business. When the organization relies on joint ventures with overseas partners, even with a strong local commitment of government funding, it still has to depend heavily on partnerships.

Organizations that pinpoint local heritage as an attraction anchorage can easily attract overseas media for collaboration and branding partnership at the beginning. In the case of publicly funded projects, substantial investment from the government attracts media spotlight in economic downturns, when art is in need of exports from Europe and the United States. Hong Kong then imports many important artworks from overseas and hence attracts criticism on the lack of investment in local art. There are differences in software infrastructure for organizations that are publicly funded or have a powerful board and patrons endowed with self-financed programs. In the long run, all organizations will have to sustain financially; however, the content of exhibitions or art activities should ideally be responsive to the public and attentive to public criticisms and the market for both publicly and privately funded organizations.

CONCLUDING OBSERVATIONS

If we look at art as a luxury item, how can it be sustained? This chapter has discussed how companies can attract and allocate social and cultural capital and employ strategies like social media and collaboration with schools to accumulate capital in the long run. Strategies also include involving key communicators who formulate a club of stakeholders or collectors, so that the market mechanism can sustain the growth of arts trade. Audience building and customer relations are also crucial in the market-driven context, embracing the force of internationalization of locally run companies. A number of successful indicators are identified in this chapter, like developing market space in similar geographical regions, employing promotional strategies and networking strategies like branding and signature events, the use of social media tools, partnering with the right institutions to broaden an audience base, and finally the operation of multi-cities strategic positioning. Table 9.1 summarizes the effectiveness of the venturing strategies employed. In the past five years these methods have proven to help enterprises be self-sufficient and independent in their financial resources. The venturing strategies provide a good starting point for future research in examining the development of cultural entrepreneurship.

Table 9.1 Application of venturing strategies for two private and two public cultural enterprises

	Private Organizations		Public Organizations	
	A	B	C	D
1. Geographical proximity/ place-making strategies	Strong	Weak	Weak	Strong
2. Promotional strategies	Strong	Weaker	Strong	Strong
3. Networking strategies	Strong	Strong	Strong	Strong
4. Partnering strategies	Strong	Strong	Strong	Strong
5. Establishment of social and cultural capital	Strong	Strong	Weak	Strong

NOTES

1. "Mind the Gap: Lessons and Findings from Engage HK," *Asia Community Ventures,* August 2013, http://www.asiacommunityventures.org/wp-content/uploads/2013/08/EngageHK_Final_Webversion.pdf (accessed December 10, 2013).
2. See http://www.gov.hk/en/about/abouthk/factsheets/docs/country_parks.pdf (accessed December 10, 2013).
3. See http://www.pps.org/reference/what_is_placemaking/ (accessed December 10, 2013).
4. "The HK Cultural Hub," *Ming Pao Weekly,* April 9, 2011.
5. *Ming Pao* and *Tai Kung Pao* are more prestigious, while *South China Morning Post* and *Hong Kong Magazine* came into the market later, followed by *Timeout* and *Artplus.*
6. Online writings include *House News* ('主場新聞', thehousenews.com), *HK InMedi*a (香港獨立媒體, inmediahk.net), and *Delta Zhi* (A 志, which is run by students).
7. Chris Bilton and Stephen Cummings (2010), 158–74.
8. See http://www.ird.gov.hk/eng/tax/ach_tgc.htm (accessed December 10, 2013).

BIBLIOGRAPHY

Bilton, Chris. 2007. *Management and Creativity: From Creative Industries to Creative Management*. London: Blackwell Publishing, 2007.

———. 2010. "Manageable Creativity." *International Journal of Cultural Policy* 16, no 3: 255–69.

Bilton, Chris, and Stephen Cummings. 2010. *Creative Strategy: Reconnecting Business and Innovation*. London: John Wiley & Sons Ltd.

Cashman, Stephen. 2010. *Thinking BIG! A Guide to Strategic Marketing Planning for Arts Organizations*. London: Arts Marketing Association and Arts Council England.

Grams, Diane, and Betty Farrell. 2008. *Entering Cultural Communities: Diversity and Change in the Nonprofit Arts*. New Brunswick, NJ: Rutgers University Press.

Hausmann, Andrea. 2010. "German Artists between Bohemian Idealism and Entrepreneurial Dynamics: Reflections on Cultural Entrepreneurship and the Need for Start-up Management." *International Journal of Arts Management* 12, no. 2: 17–29.

Leung C. C. 2014. "Collaborative Leadership between Artistic and Managing Director." *Arts Leadership in an Asian Context*. Melbourne: Melbourne University Press.

Madoff, Steven Henry. 2011. "Artist Anonymous: Anton Vidokle Disappears into the Vortex of Artworks That Is E-flux." *Modern Painters* 23, no. 3: 55–59.

Mould Oli, and Sian Joel. 2010. "Knowledge networks of 'buzz' in London's advertising industry: a social network analysis approach. "*Area (Royal Geographical Society)* 42, no. 3: 281–92.

Pow, Janette, Gayen Kaberi, Elliott Lawrie, and Raeside Robert. 2010. "Understanding complex interactions using social network analysis." *Journal of Clinical Nursing* 21: 2772–79, doi 10.1111/j.1365-2702.2011.04036.x.

Power, Dominic, and Scott Allen. 2004. *Cultural Industries and the Production of Culture*. New York: Routledge.

Preece Stephen B. 2011. "Performing Arts Entrepreneurship: Toward a Research Agenda." *Journal of Arts Management, Law, and Society* 41: 103–20.

Rensburg Deryck. 2013. "Strategic brand venturing: An intersectional idea." *Management Decision* 51, no. 1: 200–19.

Saperstein, Jeff, and Daniel Rouach. 2002. *Creating Regional Wealth in the Innovation Economy: Models, Perspectives and Best Practices*. Upper Saddle River: Prentice Hall.

Wagner, Marcus, ed. 2012. *Entrepreneurship, Innovation and Sustainability*. Sheffield: Greenleaf Publishing Limited.

Chapter 10

The Development of Cultural Entrepreneurship: Case Studies of Four Community Orchestras in Hong Kong

Chi Cheung Leung

The creative economy has been making an increasing and significant contribution to the economies in various countries, providing revenues, accounting for a percentage of the economy comparable in size to the financial services industry, and in some cases contributing more to GDP than some traditional industries (Centre for International Economics, 2009; Information Services Department, HKSAR government, 2011; Mt. Auburn Associates, 2005; Work Foundation, 2007). Many governments and enterprises have identified culture as a key resource that could help foster the development of creativity and originality (Centre for Cultural Policy Research, University of Hong Kong, 2003). To the same extent, the core of culture is creativity (Roodhouse, 2006). Cultural distinction becomes a crucial competitive advantage against the force of globalization (Mt. Auburn Associates, 2005) while cultural creativity generates exceptional value and wealth when combined with economic activities (Howkins, 2002), which benefits the economy's long-term growth and well-being (Centre for International Economics, 2009). In short, the creative economy generates economic growth and development, which fosters "income generation, job creation and export earnings while promoting social inclusion, cultural diversity and human development" (United Nations, 2010: 9).

Given the potential of the creative economy to bring thriving economic development and to foster social and cultural development, sustainable development of the creative economy becomes pertinent. Entrepreneurship is one vital element that contributes to the sustainable development of the creative industries and their economy: a critical aspect of a successful, modern society (Chong, 2002). Entrepreneurship provides the bridge between creative talents, their innovations, and the market (Aageson, 2008; Bilton and Cummings, 2010). Artists nowadays cannot afford to "turn a blind eye to the market" (Bilton, 2006: 6). They must think like entrepreneurs to survive (Mt. Auburn Associates, 2005). Creative entrepreneurs become "the catalysts that build a vibrant cultural economy" in broadening markets and creating new markets (Aageson, 2008: 95).

The increasingly competitive and fragile environment of the global economy alerts leaders of various sectors, including those in not-for-profit organizations, "to demonstrate their competency and worth, identify new opportunities for growth and innovation, and remain agile and responsive" (Wirtenberg et al., 2007: 179). Managers in arts organizations face a wide range of issues, such as greater competition for funds, diminishing government revenues and funding, reduced per-capita personal income, declining contributor population, lower program enrollment, and increasingly selective audiences (Cray and Inglis, 2011; Ott, 2001; Reider, 2001). The existing model of management and operation prompts not-for-profit organizations to better adapt to more sustainable development in the quickly changing environment. One initiative is to collaborate across sectors through strategic partnerships and to become sustainable enterprises (Wirtenberg et al. 2004). The business approach of sustainability seeks to create long-term value, embrace opportunities, and manage risks associated with various developments (Galbreath, 2011). To achieve sustainable development, not-for-profits must "operate more economically (more services for the buck), efficiently (use best practice systems and methods), and effectively (achieving maximum results with minimal resources)" (Reider, 2001: 1).

CHARACTERISTICS OF A CULTURAL ENTREPRENEUR

Cultural entrepreneurship is a critical factor for a successful modern society (Chong, 2002). According to Aageson (2008), cultural entrepreneurs are risk takers, change agents and resourceful visionaries who generate revenue from innovative and sustainable cultural enterprises that enhance livelihood and create cultural value for both

creative producers and consumers of cultural services and products ....
[They] do have common characteristics around the globe that include
being passionate, visionary, innovative, risk takers, net-workers and
leaders (96–98). They provide the bridge to connect creators and artists
to the market and consumers by using "business approaches; and deploy
financial, human and cultural capital (creativity, talent, cultural traditions,
knowledge and intellectual property) in a strategic and entrepreneurial
manner" (Aageson, 2008: 96). In short, entrepreneurs are risk-takers
(Aageson 2008; Bilton and Cummings, 2010; Chong, 2002; Kuratko and
Hodgetts, 2007) and innovators (Aageson 2008; Chong, 2002; Colbert,
2003; Rossiter et al., 2011; Reid et al., 2010) with passion and vision and
are able to act as change agents (Aageson, 2008; Kuratko and Hodgetts,
2007). They "go beyond or even against the usual ways of doing things"
They act unconventionally. For example, they produce products without
knowing in advance the demand for those products (Aageson, 2008;
Colbert, 2003). They usually think ahead (Aageson, 2008). Unfortunately,
most individuals working in the cultural sector lack the necessary financial
support, business skills, and knowledge to earn an adequate living, not to
mention to become successful cultural entrepreneurs (Aageson, 2008; Mt.
Auburn Associates, 2005). According to Kuratko and Hodgetts (2007),

> Entrepreneurship is a dynamic process of vision, change,
> and creation. It requires an application of energy and
> passion towards the creation and implementation of new
> ideas and creative solutions. Essential ingredients include
> the willingness to take calculated risks—in terms of time,
> equity, or career; the ability to formulate an effective venture
> team; the creative skill to marshal the needed resources; the
> fundamental skill of building a solid business plan; and,
> finally, the vision to recognize opportunity where others see
> chaos, contradiction, and confusion (3).

In a study (Bilton and Cummings, 2010) on the famous British
tycoon Richard Branson, the authors came up with a list of traits
that are indicators of entrepreneurial success. The list suggests that a
successful entrepreneur is usually a risk-taker, lifelong learner, swift
actor, experimenter, and thrill-seeker and a person with self-efficacy,
confidence, and experience in the industry. Bilton (2006) writes:

> Management in the cultural sector has evolved its
> own distinctive style, based on entrepreneurship, self-

management, a multi-tasking culture which challenges traditional supply chain specialization, and an economy of mixed motives which defies the classic incentive-based rationality of economics. Management in the cultural sector is essentially entrepreneurial and individualistic, dealing with risk and unpredictability by embracing diverse and apparently random processes from which coherent strategies and business models might eventually emerge (10).

In view of the above, creative managers need to be good improvisers and adapters (Bilton and Leary, 2002), play multiple roles and encourage multitasking culture within the organization, think beyond the rationale of conventional economic incentives, and be able to distinguish between sustainable development and growth. The creative industries are characterized by high levels of self-employment in which managerial and operational tasks overlap (Bilton, 2006). Among the many characteristics suggested above, seven major characteristics of a cultural entrepreneur are identified. A cultural entrepreneur is a change agent, risk-taker, innovator, multitasker, visionary leader, networker, and passionate motivator. In addition, a conceptual framework on the development of cultural entrepreneurship (Figure 10.1) is established as a theoretical base for discussion. The framework states that most not-for-profit performing groups have an artistic leader and a management team. Cultural capital, creativity, and artistic qualities are the three key aspects of endeavors, of course with different levels of attainment. The framework hypothesizes that the incorporation of a business model, the qualities of the artistic leader as a cultural entrepreneur, and the strategic planning of the management team combine to develop the cultural entity as an enterprise. These combined efforts help heighten the utilization of cultural capital, creativity, and artistic attainment with the help of innovative strategies, partnerships with businesses, and success in seeking external resources. The framework highlights that cultural capital can be utilized effectively through creativity and artistic endeavors supported strongly by reliable leadership and management. However, sustainable development requires strategic planning, effective management, visionary leadership, an innovative business model, dynamic partnership, and robust private and/or public support. All these contribute to the sustainable development of an enterprise that can generate output and revenue, which could impact the cultural industries and subsequently the economy at large. In other words, cultural entrepreneurship incorporates business strategies in running a cultural entity that synergizes original, artistic, and innovative creativity.

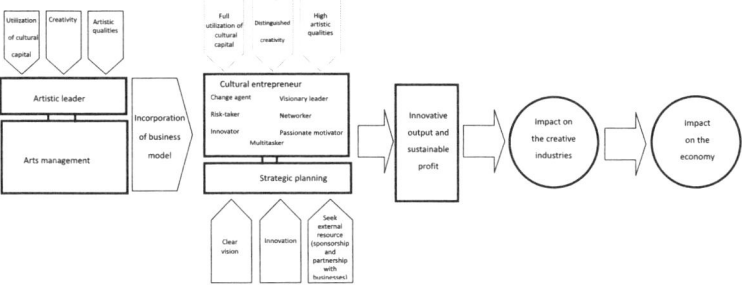

Figure 10.1 Development of cultural entrepreneurship

BACKGROUND AND THE STUDY

Hong Kong has many community Chinese orchestras and ensembles. They usually have one annual concert with some infrequent performances. Most of them are not-for-profit organizations, and players do not receive payment for their performances. A very small number of them pay an honorarium to players for rehearsals and performances. The 1970s and 1980s was the golden era of community Chinese orchestras in Hong Kong, with many active orchestras. But with the 1977 establishment of the Music Office (MO) (音樂事務處), which offers low-tuition instrumental training to selected students and forms orchestras in different district centers; the 1978 formation of the Hong Kong Chinese Orchestra (HKCO) (香港中樂團), the first and only professional Chinese orchestra in Hong Kong; the training of professional Chinese instrumentalists by the Hong Kong Academy of Performing Arts (香港演藝學院) in 1984; and the inflow of professionally trained musicians from mainland China, the prosperous era of Chinese orchestras began to gradually wane. Many Chinese orchestras in Hong Kong started to face challenges from high-quality performances by HKCO, MO orchestras, and professional musicians. Despite all these challenges, not-for-profit Chinese orchestras continue, but they do not lead in the field as they did in its golden era.

Community performing groups are valuable social capital that benefits the community (Langston and Barrett, 2008; Veblen and Olsson, 2002). This cultural or social capital includes many freelance professional musicians, emerging artists, talented adolescents, experienced amateur musicians, and retired people with musical talents. Building a community orchestra with continuous development into entrepreneurship maximizes available cultural capital from the community by engaging people in valuable, inexpensive, and productive cultural and creative activities. In particular, "our population will live more and more productive years after retirement" and "the importance of volunteers for nonprofits should continue to rise over the upcoming decades" (Ott, 2001: 316). The effective utilization of this huge amount of capital benefits audiences, musicians, and the community at large. One of the most obvious reasons for supporting a community performing group is to promote musical culture in the community, which includes educating the public, engaging amateur musicians in musical performances, facilitating the continuation of musical lives beyond formal schooling, providing informal adult education (Mark, 1992), and maximizing available cultural capital.

Orchestras in general share two characteristics in management—operational and artistic. While the executive director or executive committee takes care of day-to-day administrative matters and operational decision making, the music director is in charge of artistic decision making on program matters, auditioning and selection of players, and choices of guest performers and soloists. But their work may sometimes overlap. The executive director and artistic director need good communication and cooperation, particularly when decisions relate to external factors (Bilton, 2006; Cray and Inglis, 2011). The overlapping responsibilities between operational and artistic matters are particularly distinctive in not-for-profit orchestras. Another distinctive feature of a not-for-profit organization is the engagement of volunteers (Hussey and Perrin, 2003; McCurley, 2005; Ott, 2001). They are unpaid and motivated primarily by their love for music, with little ulterior consideration (Drinker, 1992; Drucker, 1990). However, their quality of performance may not be at the desired levels (Reider, 2001). Furthermore, huge efforts are required to keep the flame of the volunteers alive—not allowing their commitment to become just a job. Thus the ability of the management team to attract and hold committed people becomes a decisive factor in the success or failure of an organization (Drucker, 1990). Effective leadership and management are thus crucial in sustaining players' passion in music and their dedication to the orchestra.

This study looks into the development of a group of not-for-profit community Chinese orchestras in Hong Kong, aiming to identify key characteristics in management and operational practices and to present a proposal on how cultural entrepreneurship could be embedded in the development of not-for-profit orchestras in accordance with a theoretical framework (see Figure 10.1) built on the reviewed literature. With reference to the framework, a self-assessment questionnaire was established and conducted among key leading figures (mainly music directors) in the orchestras. The survey was in two parts; part one asked leaders to evaluate themselves as cultural entrepreneurs: as change agents, risk-takers, innovators, multitaskers, visionary leaders, networkers, and passionate motivators; part two evaluated the extent of the development of orchestras toward entrepreneurship based on key items chosen from the theoretical framework. The survey served as preliminary reference data for follow-up phone interviews. Questions were asked in accordance with the results of the survey. Four not-for-profit Chinese orchestras and their artistic leaders were chosen for the study. The orchestras were founded across four decades, respectively. They are Lok Sum Chinese Orchestra (樂心中樂團) (founded 2008), Hong Kong Taoist Orchestra (香港道樂團) (founded 1996), New Tune Chinese Orchestra (新聲國樂團) (founded 1987), and Yao Yueh Chinese Music Association (樂樂國樂團) (founded 1974). The last orchestra was founded by the author, who is currently its music director. The study of this orchestra was thus conducted from the perspective of a participant-observer with an inside-out/outside-in understanding of the orchestra. An in-depth study of the orchestra was done in more detail through a separate internal survey carried out earlier. The findings of the study are discussed with reference to the established framework in order to shed light on the conceivable development of not-for-profit orchestras toward entrepreneurship. The results give insights of cultural entrepreneurs-to-be, practitioners in the cultural industries, and policy makers on the sustainable development of performing groups, in particular nonprofits.

LOK SUM CHINESE ORCHESTRA

Lok Sum Chinese Orchestra (LSCO), one of the youngest Chinese orchestras in Hong Kong, was established in 2008. It consists of around thirty players, who were initially recruited through social media, such as forum sites and Facebook. The chairperson, Ho Man Chun (何文津), who plays dizi (Chinese flute) and flute, is an energetic young man, passionate about promoting Chinese music. In the self-assessment survey, Ho was confident of himself as a multitasker, a visionary leader, and a social networker. As a multitasker, he is involved in various tasks in public relations, sponsorship, rentals, stage work, concert programming, seating plans, rehearsal schedules, finance, and internal communication. Ho states, "Amateur Chinese orchestras [in Hong Kong] have a common feature ... [they] perform very similar repertoire. ... [LSCO] wants to be innovative ... and attract audiences that have never heard Chinese music before." He continues, "Currently most Chinese music concerts depend on the support of friends and relatives, but we want to open up the market to new audiences, through performing popularly known music hoping that ... they would finally appreciate traditional Chinese music." He is critical about the situation of community Chinese orchestras but hopeful about attracting audiences beyond friends and relatives. As a social networker, he is proactive in connecting musicians and conductors, seeking advice from professors, and searching for external resources and sponsorship.

Apart from playing traditional Chinese music, LSCO plays pop music, cartoon theme songs, computer game music, and jazz. Members of the orchestra write or arrange music that appeals to young people. Sometimes members incorporate multimedia elements into their performances. More than half of members play both Chinese and Western instruments. When the orchestra does not have the specified Chinese instruments, performers replace those instruments with Western ones. LSCO aims to enrich traditional Chinese music by providing audiences with a refreshing musical experience. In addition to creativity, the orchestra is well aware of the importance of artistic quality. Newly composed or arranged works have to be assessed and approved by the head of the composing team, while two elite groups selected from among the membership perform traditional Chinese music and jazz, respectively. In addition to holding concert performances in shopping malls and ordinary venues, LSCO also plans to perform in streets, a trend of the younger generation of artists. Further, they promote their music by uploading performances onto YouTube; some performances have been viewed thirty thousand to forty thousand times. LSCO is self-supporting in management and financial matters. Members have to pay regular fees to pay the expenses of the orchestra, including renting a rehearsal space in an industrial district. LSCO also finds revenue by contracting a public relations company to increase performance opportunities.

LSCO has a few features of cultural entrepreneurship, even though members are young. It is visionary in targeting Chinese music audiences of the younger generation; using commercial businesses like the public relations company to increase revenues; recruiting members with composing and performing ability; forming small ensembles comprised of elite players to ensure creative and artistic quality; and utilizing social media to seek external attention and collaboration opportunities. Such an orchestra formed via social media, though not typical of Chinese orchestras in Hong Kong, may become common among young artists. On the one hand, the sustainability of this kind of orchestra could be very challenging in view of the financial burden these young people have to bear. On the other hand, the new and innovative ideas of these young people cannot be underestimated, particularly with the design of the elite groups and the adaptation of the business model.

HONG KONG TAOIST ORCHESTRA

The Hong Kong Taoist Orchestra (HKTO) was established in 1996 as a not-for-profit entity under the financial support of Fung Ying Seen Koon (FYSK) (蓬瀛仙館), a Taoist religious organization. The chairperson and artistic director and of HKTO, Liu Hong (劉紅), is a professor at the Shanghai Conservatory of Music. He has had intensive musical training at Wuhan Conservatory of Music and Chinese University of Hong Kong. The orchestra has around forty players recruited via its website. It offers free instrumental classes to attract people to join the orchestra. Taoist music is usually played at traditional events such as funerals, rituals, and banquets. People don't usually perceive Taoist music as a performing art. Some may consider ritual music less artistic and of lower social status. The public has the perception that Taoist priests performed music for money; at the same time, movies and television shows portray a negative image of Taoist priests. According to Lau, by performing Taoist music on stage, the orchestra portrays a positive image to the public. The purpose is to improve the social status of Taoism.

HKTO and the Chinese Taoist Association organize the Annual Taoist Music Festival (年度道教音樂匯演), inviting Taoist orchestras from various places to perform on stage. The festival is held in different cities each year and has become an important cultural event of the Taoist religion. Taoist music was originally transmitted orally and aurally. To make the music more accessible, members of HKTO have started transcribing the music in written notation. Lau states that he utilizes cultural capital to a vast extent: "I apply my research in [operating] the orchestra. These are [cultural] resources. As minute as an arrangement, I study the source through research." With regard to its impact on creative industries, Lau argues, "HKTO has no connection with commercial activities because FYSK is a charity organization. But in the aspect of culture ... we have motivated the formation of Taoist orchestras for performance purpose. As such the impact is intense."

Of the seven features listed in the survey, Lau says he performs best as an innovator, multitasker, and motivator. In terms of innovation, Lau claims that HKTO is the first of its kind in transforming ritual Taoist music into a performing art in Hong Kong. As a multitasker, Lau is a researcher, arranger, conductor, coach, music director, organizer, cultural promoter, and administrator. As a motivator and social networker, Lau is successful in recruiting nonbelievers of Taoism to join HKTO and in organizing the Annual Taoist Music Festival since 1996. The festival has attracted Taoist orchestras from different parts of China. As a change agent, risk-taker, and visionary leader, he is constrained by the nature of Taoist music, which has deep cultural context. He says, "First, I think I cannot go too far [artistically] because audiences may query the difference between Chinese music and Taoist music. Second, as an amateur group, the management [model] cannot be too far away from the tradition [nonprofit making]." He continues, "What I am now conducting is a cultural activity with long history. There is not much room for change. From my perspective, it is already visionary to allow some people to understand Taoist music."

HKTO is a unique example of a Chinese orchestra in Hong Kong that is strongly linked to Taoism, a traditional Chinese religion. The impact of creative industries on the economy might be tiny, but that on culture can be long lasting. With stable financial support from FYSK and a professionally trained artistic director, HKTO has a strong foundation to develop further as a cultural enterprise, not in the sense of profit making but in cultural capital building. This is an affiliation-based model whereby the orchestra has substantial support—particularly financial support and a rehearsal venue provided by the host organization. Its development is more stable than that of other community orchestras, which depend heavily on the financial support of members, sponsors, donations, and ticket sales.

NEW TUNE CHINESE ORCHESTRA

Established in 1987, New Tune Chinese Orchestra (NTCO) is a large enterprise, with nearly three hundred members and players. The leading figure of this orchestra is its music director and conductor, Yau Siu Pun (邱少彬), who plays several Chinese instruments and teaches these instruments in schools. Players in NTCO are his students. Most of them join NTCO when they are at school and continue to stay after graduation. Now they work in different professions, and around fifteen of them have become instrumental tutors under Yau's supervision. NTCO is a highly structured organization, with eleven departments taking care of education, examinations, external affairs, publicity, reception, information and publications, logistical support and general affairs, scores, membership, volunteer services, and a friends club. NTCO organizes various performances, including joint school concerts, charity concerts, performances in mainland China and Taiwan, and other music activities, such as instrumental classes, competitions, graded examinations, symposiums, and seminars. Its repertoire ranges from traditional works and arrangements to new original works. NTCO also promotes rarely performed works and new works by Yau and his students. It has frequent collaborations with musicians from mainland China and Taiwan. Furthermore, the orchestra publishes newsletters, symposium proceedings, and concert CDs. Its achievement has been recognized and recorded in an almanac of Chinese culture, and it has received an honorary cultural award.

In the self-assessment survey, Yau rated himself high in all traits of a cultural entrepreneur, except as a risk-taker. He explains, "Taking risk means a one-way relationship but I have to evaluate the circumstances in order to plan. I have no chance to take risk because of objective constraints." In a phone interview, Yau said that he is a change agent who pursues artistic improvement and high standards of performance. As an innovator, he highlights his creativity in writing new compositions. He states, "Without sustainability [in creativity], it is impossible to become mature [in artistic pursuit]." As a multitasker, he is a mentor, administrator, organizer, motivator, and communicator. As a visionary leader, Yau is more rational: "I hope we can improve step-by-step to meet the standard of a professional orchestra." As a social networker, Yau is chair of the Kwai Chung and Tsing Yi District Culture and Arts Co-ordinating Association (葵涌及青衣區文藝協進會), and NTCO is able to rehearse free-of-charge in the Kwai Fong Community Hall (葵芳社區會堂). Yau also has very good relationships with musicians in mainland China, which provides performance opportunities on the mainland and access to Chinese music scores that are not widely published. As a motivator, he is successful in motivating his students to run NTCO and to teach instrumental music in the community.

Yau has no intention of changing NTCO into a cultural enterprise using a commercial model because he is well aware that the orchestra cannot make profits. He says, "Even with much better planning, the quality of the musicians and the time they can afford is still far from feasible." In view of its size, structure, and activities, NTCO is surely a cultural enterprise but not one that can become profit making. It follows a typical mentor–mentee model, whereby the music director is also the instrumental tutor of the orchestra members. Its sustainability depends very much on a succession plan, with the mentor facilitating his mentees to take up his duties and leadership role.

YAO YUEH CHINESE MUSIC ASSOCIATION AND ITS ORCHESTRA

Yao Yueh Chinese Music Association (YYCMA) was founded in 1974 as a not-for-profit organization by a group of high school graduates with a vision of promoting Chinese music. This is one of very few long-standing orchestras that survives and continues to grow. In 2007 YYCMA started to reengage itself proactively in an attempt to reposition the orchestra from an entrepreneurial perspective. The question is whether changing this cultural asset is feasible given the many constraints and weaknesses community orchestras are facing.

The mean age of orchestra members is thirty-four (with individuals ranging from thirteen to seventy-four years old), with 40 percent of them age twenty-one to thirty. Most of them are secondary school or tertiary institute students (32.5 percent), administrators or white-collar workers (26.3 percent), or teachers (19.3 percent). Others include a nurse, a firefighter, retirees, professional musicians, and housewives. Forty-two percent indicate that they are engaged in part-time jobs related to music teaching or music performance. Almost half of the respondents (46 percent) have received some kind of musical training. The rest received their training from private lessons, music classes in secondary school, the Music Office, and other sources. Many of them are experienced musicians, with a mean length of playing their instruments of 15.7 years. Nearly 60 percent of them have played their instruments for over ten years, while a quarter of them have six to ten years of experience. Around 40 percent of them are newly joined players (three years or less with the orchestra), and the rest have been in the orchestra for over three years.

Unlike established professional orchestras, YYCMA has neither a board of directors nor a council—let alone a team of full-time administrators—to take care of the management and daily administration of the orchestra. The operation of the orchestra is mainly overseen by an executive committee. Committee members are elected by registered members for terms of two years at a general assembly meeting. The elected committee appoints the music director of the orchestra, and the music director takes care of the artistic directions of the orchestra. The executive committee also appoints an organizing committee to take charge of operational duties for each concert. The executive committee decides the dates, venues, programs, and rehearsal arrangements for concerts, while the organizing committee concentrates mainly on ticket sales, marketing, promotion, and the printing of posters and concert programs. Part-time workers may be employed to assist with stage work, transportation, photo taking, video shooting, audio recording, and other logistical work when funds are available. If not, musicians have to play dual roles, taking care of logistics as well as performing in concerts on a voluntary basis. So far, nearly half of the musicians have been executive committee members and/or organizing committee members. In a recent effort to celebrate the fortieth anniversary of YYCMA, a committee was formed among social leaders and senior members to extend the orchestra's social connections.

The music director, Leung Chi Cheung (梁志鏘), determines the artistic direction of the orchestra. He is also the founding conductor of the orchestra. In the self-assessment survey, Leung said that he has taken on a multitasking role, and very often his duties overlap those of the committees. These duties include recruitment of players, program design, choice of soloists, rehearsals and rehearsal scheduling, promotion and marketing, and even the details of printing and stage work. Among these, the recruitment of new players has been a major challenge. He also considers himself a change agent, an innovator, a visionary leader, a social networker, and a motivator, as reflected in the orchestra's achievements in recent years. As a risk-taker, he has reservations, but he admits that he takes calculated risks from time to time on artistic and strategic matters. For example, while getting enough players to form a full orchestra is a great challenge, getting enough quality players is an even greater challenge. The strategy regarding recruitment of musicians is to gather enough players first before improving standards gradually. Unlike other orchestras, no audition is required; anyone who wants to join the orchestra is welcome. To a certain extent, this is a risky approach, because conducting an orchestra with members having diverse skills is difficult. Musicians are often meticulous and demanding but are also delicate and insecure. Thus the music director must have the ability and confidence to train musicians and must be sensitive to their psychological needs and technical standards. Nonetheless, maintaining high rehearsal and performance standards is important in managing a large group of musicians. Insisting on quality and taking into account musicians' varied technical skills involves a calculated risk. Leung believes the best way for a performing group to improve is through more performance opportunities. With specific performing targets ahead, the director can push and encourage musicians to achieve higher standards. Players are given chances to improve from one concert to another, and performance opportunities can serve to attract new musicians. In this way, the orchestra is able to increase the number of musicians and improve quality. The number of musicians increased from fewer than twenty in 2007 to more than seventy to date. On average, the orchestra offers three different programs annually, compared to usually one annual program in the past. Upcoming performances will highlight some elite young artists as soloists and later will feature them as members of small ensemble groups to underscore their talent.

As a participant-observer, the music director evaluates the orchestra over the past five years as being above average in the utilization of cultural capital, creativity, artistic endeavors, leadership, vision, innovation, seeking external resources, and incorporating a business model. YYCMA has extensive connections in the artistic field, which facilitates collaboration with renowned soloists and artists at major concerts. The orchestra emphasizes the engagement of local artists and original works written by Hong Kong composers. Other than the programming of orchestral works, the orchestra has branched out into two major types of programs: choral music and ensemble/chamber works written by Hong Kong composers. YYCMA has launched two major choral work concerts: one featuring Chinese history and one featuring Chinese poetry, which has given the orchestra a good reputation for choral music accompanied by a Chinese orchestra. As for ensemble/chamber works, two concerts featuring works written by Hong Kong composers were complemented with traditional Chinese music. Presenting both traditional and contemporary works in a concert helps to broaden, vary, and balance the taste of audiences from traditional to contemporary and vice versa. The choral music programs highlight the orchestra's strength in conducting a large-scale concert while the ensemble/chamber programs refresh the audience by underscoring contemporary works, particularly those written by Hong Kong composers.

The concert focusing on Chinese history received a press review. The concert combined calligraphy, poems, spoken verse, historical background, painting, and dance, with the various art forms stimulating and enriching the imagination of audiences in appreciation of the music performed. Below are some statements from the review of the concert:

The orchestra was confident and musical.

- The well-balanced choir and orchestra work together incredibly well, gaining a depth of melancholy, lusciousness and emotional performance.

- The orchestra's calm performance and strong coherence of musical style has delighted the audience.

- The excellent and creative combination of art forms during the concert has brought the concert into a great success... allowing innovation and intellectuality to shine forth.

- The concert is overwhelmingly enjoyable. ... The audiences left feeling fulfilled, and entirely spent (Leung, 2009).

The concert embraced the audience with different art forms in the context of major Chinese historical events, which enlightened and heightened audience appreciation of the music performed.

In the past, YYCMA performed one concert each year. Ticket sales were mainly promoted internally by the players to their friends and relatives. With the increase in the number of concerts each year, the orchestra can no longer depend solely on a single mode of promotion. Publicity has to be done in multiple modes. Yet, due to the limited budget for each concert, publicity must be done at a low cost. One way to make low-cost publicity possible is to adopt methods that benefit not only the orchestra but also parties that help the orchestra. This arrangement helps achieve win-win promotion outcomes, where the orchestra receives free publicity. The orchestra needs to reach audiences of different generations and tastes. Other than the traditional tactics of placing pamphlets and posters at various performance venues and mailing them out to schools, professional music groups, and music studios, the orchestra also approaches the media, offering attractive stories to the press about the orchestra and its musicians. A few articles on the orchestra were featured in both English and Chinese newspapers. In one of the featured articles, free tickets were given out to readers who could answer questions correctly. The orchestra offered to perform free on radio and television programs in exchange for interviews with featured artists, successfully promoting concerts. Besides traditional media, technology allows the orchestra to reach audiences through updates and news about the orchestra, its artists, and rehearsals on its website and various Internet sites like Facebook. Most of this publicity (except the printing of pamphlets and posters) does not impose an extra financial burden. With a limited budget, the orchestra is always asking for help from volunteers, mostly from members. Furthermore, discount group tickets are offered to students, and sometimes free tickets are given to seniors and underprivileged families. With these multiple methods of publicity, the orchestra is able to maintain a fairly good attendance rate at concerts, varying from 60 percent to a full house, with some minor revenue.

Though YYCMA hasn't had any business model to follow, it has started to partner with businesses. One partnership was initiated by a commercial music studio and instrument company. This initiative is an example of the orchestra's adoption of mutually beneficial collaboration. The idea is to have registered members of the orchestra teach in the studio while the studio branches out its business in providing Chinese instrumental music lessons. Members of the orchestra get income while the company donates part of its profit to the orchestra. This new venture is still in an experimental stage; its actual effectiveness will be evaluated after its full implementation. The mutually beneficial nature of this partnership initiative motivates both parties to contribute. Trust and persistent cooperation between the two parties and quality lessons surely are crucial to the success of this initiative.

Another partnership initiative is the launching of music gatherings for musicians and the audience. The gatherings are partially paid for by a donor. Musicians and audience members meet at a boatyard not far from the city. The boatyard waterfront can accommodate sixty people at a time. Food and drinks are provided to participants at a reasonable charge. The initiative allows musicians of the orchestra to share their music with audience members in a relaxed environment and allows the audience and musicians to have further conversations after performances. This is an audience-building initiative. It also aims to attract donors and build partnerships for the orchestra. Communication between the audience or possible donors and the musicians is necessary, as it provides a chance for orchestra members to better understand the expectations of donors and the audience. This is an outreach initiative that allows members of the orchestra to think beyond music and to help build the image of the orchestra through communication with people outside the music profession. This initiative is more than a friends club; it is a gathering that helps promote Chinese music and serves as a platform for audience building and the gathering of donations for the development of the orchestra. Though professional orchestras commonly conduct these kinds of activities to attract donors, it is not usual for amateur Chinese orchestras to launch similar activities.

YYCMA is proactive in attracting various kinds of public funding. Most of the funding received is given on a project basis. For each concert project, the orchestra submits a proposal to be assessed by music professionals assigned by the funding body. The orchestra has been successful in bidding a number of grants from public funding organizations, competing against proposals submitted by professional groups. However, it is hard for the orchestra to sustain its development on project-based grants. Until the orchestra receives a long-term grant, it will be unlikely to sustain itself with long-term plans. With this understanding, the orchestra has been actively searching for private funding support. The major strategy is to show the public and potential donors that the orchestra has a long-term vision and achievable goals. To achieve this, YYCMA plans innovative programs and describes the concert series in a brochure to attract donors. Other than classic repertoire in Chinese music, the concert series integrates micro-movies, a laptop-computer orchestra, and digital arts with Chinese music. It also highlights young musicians. In addition, the program invites collaborations with major soloists, conductors, and other Chinese orchestras. With these achievements and social connections, the orchestra has been able to attract donations to support basic operations for three years. This also forms a base for seeking more donations. For the orchestra, this is a breakthrough. The next step will be more demanding and challenging. To pave the way for sustainable development, the enhancement of both artistic and operational quality for the orchestra is inevitable. Planning, concert programming, quality artistic and creativity endeavors, management, administration, and operation are crucial areas for the orchestra to advance.

The orchestra has survived for nearly forty years with experienced musicians—many of them working as freelancers. In recent years, the orchestra has made a gradual shift from being a not-for-profit performing group toward being an active community orchestra with more frequent performances; innovative music programs; a repertoire including choral works and new works by Hong Kong composers; participation by local professional musicians; external sponsorships; marketing strategies; publicity via the press, television, and radio; and partnerships with commercial organizations. The persistent, assertive, and visionary management of the orchestra has led to the enhancement of its public image and member confidence. But it is still far from attaining cultural entrepreneurship with profitable income.

All the projects mentioned above have been managed through the voluntary work of member musicians. With the rapid development of the orchestra, members have had to take up duties that in a professional orchestra are performed by a full-time administrative team. The capacity of volunteers to handle the strategy initiatives and daily operation is reaching its limit. The quality of management depends heavily on enthusiastic volunteers. At the same time, the demand for quality orchestra performance continues. Both the standard of performances and innovations in programs are vital in attracting audiences. The increase in the number of concerts poses higher demands for volunteer work from musicians in the areas of operation and management. Thus both artistic quality and operational quality are challenges. The question is how much more energy and time can the volunteers spend on running the business of the orchestra? Can a business model of management help ensure quality implementation? Change and transition will be inevitable if this cultural entity continues toward entrepreneurship. The major issue is whether it is going the right way. Even if it is taking the right direction, how and in what ways should it change to achieve high artistic creativity and commercial products that generate income? The way forward for YYCMA is challenging.

LIMITATIONS OF CULTURAL ENTREPRENEURSHIP

The four leaders of the Chinese orchestras have different qualities of cultural entrepreneurs: they are change agents, risk-takers, innovators, multitaskers, visionary leaders, networkers, and passionate motivators to varying extents. They are passionate leaders who have been committed to the development of their orchestras for many years. All four have cultural entrepreneurial qualities but lack some expertise. At the same time, the leaders are aware of the constraints that inhibit the development of their orchestras. Not all the leaders intend to move forward as cultural entrepreneurs, particularly in the sense of making a profit, but they do want their orchestras to be sustainable in their development. Each orchestra has its own way of survival and is eager to enhance its artistic standards. As not-for-profit groups, their sustainability depends very much on the passion of volunteers and on the leadership of their artistic directors. Further advancement will not be easy, as the standard of players varies and the resources they can acquire are limited. Moreover, these cultural entities do not produce profits that can sustain their development, and their impact on the creativity industries and economy is minimal in terms of GDP, although their potential cultural capital has been underestimated. Table 10.1 provides a summary of the features of the four Chinese orchestras:

Table 10.1 Features of four community Chinese orchestras in Hong Kong

	New-Generation Model	Affiliation-Based Model	Mentor–Mentee Model	Developmental Model
	Lok Sum Chinese Orchestra	HK Taoist Orchestra	New Tune Chinese Orchestra	Yao Yueh Chinese Music Association
Founding year	2008	1996	1987	1974
Membership	30	40	300	70
Initiation	By a high school student via social media	By a religious organization	By the music director, who is also mentor of the players	By a group of high school graduates
Financial status	Main source of support through membership fees and member donations	Sponsored by a religious organization with a free rehearsal venue	Self-supportive with a free rehearsal venue (some members earn income through tuition)	Self-supportive with a free rehearsal venue; income mainly from public funding, private donations, and ticket sales
Utilization of cultural capital	Mainly members and some friends integrating multimedia aspects into Chinese music	Knowledge transfer of research on a deep-rooted culture to music performance	Able to mobilize three hundred amateur musicians for various performances, plus collaboration with choirs, dance troupes, etc.	Integrating Chinese music performance with various cultural elements like chorus, calligraphy, history, visual arts, digital arts, computer music, and dance
Creativity and artistic qualities	Members compose and arrange music that appeals to the young generation	The artistic director studies, transcribes, and writes Taoist music for performance	Performs works rarely performed or those written by the music director	Emphasizes the performance of works written by Hong Kong composers

Incorporation of business model	Partners with public relations company for outreach performances	No need for partnership as it has stable financial support from its sponsor	Self-supportive, with income from tuition, graded examination fees, etc.	Proactive in seeking external funding and partnership with a commercial studio for provision of tuition
Management and strategic planning	Management under a group of elected members with innovative ideas	Managed by the board of directors and involved in the Annual Taoist Music Festival	Highly structured organization with eleven departments dealing with operations of the various businesses of the orchestra	Managed by an elected committee with external members with social connections
Innovation	Insightful targeting of a young audience	Successful in promoting Taoism to non-Taoists using Taoist music performance	Able to sustain a performing group for twenty-six years with leadership by a single music director	Innovative in publicity, integrating multimedia elements into Chinese music performance, and seeking external support
Innovative output and sustainable profit; impact on creative industries and the economy	The orchestras do not have innovative output that could generate sustainable income, and their impact on creative industries and the economy is minute from the perspective of GDP. However, it should be noted that their impact on culture and their potential capital that could contribute to culture have not been estimated or have been underestimated			

The four orchestras use different strategies and adopt diverse approaches to outreach and to promoting Chinese music. LSCO is energetic and innovative. It belongs to the new-generation model, which is highly mobile and appeals to young audiences. HKTO is affiliated with a religious group. The direction of its development depends very much on the sponsoring body. The artistic director is critical to its artistic quality, but if the sponsoring body were not a religious group, the constraints on its artistic direction would be less. NTCO is a typical mentor–mentee model that is quite common in Hong Kong. Surely this is a successful model that showcases the persistence of the music director in leading the orchestra. YYCMA's music director seeks to adapt to the changing cultural scene in Hong Kong to better sustain the orchestra in a different perspective. Nonetheless, the model needs a high level of commitment and team leadership. The major features of the four community orchestra models are categorized in Table 10.2.

Table 10.2 Features of four community orchestra models

Model	Features
New-generation model	Innovative and trendy in its creative and artistic approach, which appeals to young audiences Independent in its administration and flexible in seeking solutions
Affiliation-based model	Affiliated with a sponsoring body that provides stable administrative and financial support Depending on its nature, the sponsoring body may or may not affect the artistic and developmental direction of the orchestra.
Mentor–mentee model	The artistic leader is the mentor of players in the orchestra, providing a strong mentor–mentee relationship. The sustainability of the orchestra depends very much on the leadership of the mentor and the succession plan, if any.
Developmental model	The orchestra is assertive in making changes to adapt to new developments. Demands strong team leadership and a commitment to fulfill visionary plans

From consolidation of quantity and quality of musicians, brand building through program design, and innovative marketing strategies in audience building to new ventures and partnerships with commercial enterprises, these orchestras have passed through a long journey toward something promising but also uncertain: cultural entrepreneurship. The traditional model of most professional orchestras heavily relies on donations and sponsorship, areas that are facing serious financial turmoil. The four community orchestra models in this study distinguish themselves with various features. In the pursuit of sustainability, they use effective methods that fit their needs and their markets. Cultural entrepreneurship can be approached in different ways, but not much has been studied in this area. The concept of cultural entrepreneurship is much geared toward the production of profits and the increase of sustainable revenue. The cultural value and high potential of cultural capital in the community—in this case community orchestras—could have a huge impact on quality of life, but they have not been fully deliberated. Surely this is an important issue for discussion in future studies.

Under the concept of cultural entrepreneurship, a community orchestra can form partnerships with commercial and noncommercial entities in running innovative programs that can support development of the orchestra. To achieve this, the orchestra can reach out to promote its own brand, mission, and concert programs to possible donors and simultaneously to audiences. Neither the artistic nor operational aspects can be neglected. The orchestra needs a core team of passionate people working together for its development—skilled musicians on the artistic aspects, and professionals on the management and fund-raising aspects. They need to work hand in hand with a common goal and vision for the orchestra's future. Of course, success depends very much on detailed planning, operations, and harmonious cooperation among musicians and administrators. Leadership and management in drawing people together for a common goal are fundamental.

Moving from not-for-profit status to entrepreneurship, from volunteerism to paid jobs, from amateur to professional status, from single musicians to partnerships, and from donations and sponsorships to a business model—all of this is challenging. How far can the orchestra build up its social connections and attract donors? How much donation funding is enough? Can the orchestra really become a cultural enterprise? Can it achieve sustainable development and make a profit in the long run? Would the orchestra go back to the traditional model of relying heavily on donations? Is entrepreneurship realistic for a community orchestra? What should be the way forward? The concept of cultural entrepreneurship poses a lot of risks and uncertainties; careful studies on its possibilities are well worthwhile, and policy makers and managers of cultural entities should not overlook the potential benefits it will bring. There are a lot of opportunities for community performing groups. They are innovative and promising but need the support of the right people and policies to grow and flourish.

Note: This article is reprinted with permission from Emerald Group Publishing Limited under Leung, C. C. (2013). *The Development of Cultural Entrepreneurship: Case Studies of Four Community Orchestras in Hong Kong. Asian Education and Development Studies* (AEDS), 2 (3), 275-293.

BIBLIOGRAPHY

Aageson, Thomas H. 2008. "Cultural Entrepreneurs: Producing Cultural Value and Wealth." In *The Cultural Economy*, edited by Helmut K Anheier and Yudhishthir Raj Isar, pp. 93–107. Thousand Oaks, CA: Sage.

Bilton, Chris. 2006. "Cultures of Management: Cultural Policy, Cultural Management and Creative Organisations." Draft paper for the seminar "Management of Culture/Culture of Management," June 28, 2006. Coventry, UK: Warwick Business School.

Bilton, Chris, and Stephen Cummings. 2010. *Creative Strategy: Reconnecting Business and Innovation*. Chichester, UK: Wiley.

Bilton, Chris, and Ruth Leary. 2002. "What Can Managers Do for Creativity? Brokering Creativity in the Creative Industries." *International Journal of Cultural Policy* 8, no. 1: 49–64.

Centre for Cultural Policy Research. 2003. "Baseline Study on Hong Kong's Creative Industries for the Central Policy Unit, HK Special Administrative Region Government." Centre for Cultural Policy Research, University of Hong Kong, http://www.cpu.gov.hk/english/documents/new/press/baseline%20study(eng).pdf (accessed August 25, 2010).

Centre for International Economics. 2009. "Creative Industries Economic Analysis, Final Report," Centre for International Economics, http://www.enterpriseconnect.gov.au/who/creative/Documents/Economic%20Analysis_Creative%20Industries.pdf (accessed December 16, 2011).

Chong, Derrick. 2002. *Arts Management*. London: Routledge.

Colbert, François. 2003. "Entrepreneurship and Leadership in Marketing the Arts." *International Journal of Arts Management* 6, no. 1: 30–39, http://www.gestiondesarts.com/fileadmin/templates/main/PDF_Publications/IJAM/V61_Colbert.pdf (accessed November 15, 2011).

Cray, David, and Loretta Inglis. 2011. "Strategic Decision Making in Arts Organizations." *Journal of Arts Management, Law, and Society* 41: 84–102. http://www.tandfonline.com/doi/abs/10.1080/10632921.2011.573444 (accessed October 28, 2011).

Drinker, H. S. 1992. "Amateurs and Music." In *The Music Educator and Community Music*, edited by M. L. Mark , pp. 36–39. Reston, VA: Music Educators National Conference.

Drucker, Peter F. 1990. *Managing the Non-Profit Organization: Principles and Practices*. New York: HarperCollins Publishers.

Galbreath, Jeremy. 2011. "Strategy in a World of Sustainability: A Developmental Framework." In *Handbook of Corporate Sustainability: Frameworks, Strategies and Tools*, edited by M. A. Quaddus and M. A. B. Siddique, pp. 37–56. Northampton, UK: Edward Elgar Publishing.

Howkins, John. 2002. *The Creative Economy: How People Make Money from Ideas*. London: Penguin.

Hussey, David, and Robert Perrin. 2002. *How to Manage a Voluntary Organization: The Essential Guide for the Not-for-Profit Sector*. London: Kogan Page.

Information Services Department, HKSAR government. 2011. "Hong Kong: The Facts," http://www.gov.hk/en/about/abouthk/factsheets/docs/creative_industries.pdf (accessed December 14, 2011).

Kuratko, Donald F. 2007. "Entrepreneurial Leadership in the Twenty-First Century." *Journal of Leadership and Organizational Studies* 13, no. 4: 1–11, http://search.proquest.com/docview/203134024?accountid=11441 (accessed November 15, 2011).

Kuratko, Donald F., and Richard M. Hodgetts. 2007. *Entrepreneurship: A Contemporary Approach*. 7th ed. London: Thomson/South-Western.

Langston, Thomas W., and Margaret S. Barrett. 2008. "Capitalizing on Community Music: A Case Study of the Manifestation of Social Capital in a Community Choir." *Research Studies in Music Education* 30, no. 2: 118–38, http://rsm.sagepub.com/content/30/2/118.full.pdf+html (accessed August 15, 2011).

Leung, M. K. 2009. "Let the past gone with music—Review on Yao Yueh Chinese Music Orchestra *Reminiscence of the Red Cliff* Concert." *Wenweipo*, May 17, C2.

Mark, Michael L., ed. 1992. *The Music Educator and Community Music*. Reston, VA: Music Educators National Conference.

McCurley, Steven. 2005. "Keeping the Community Involved: Recruiting and Retaining Volunteers." In *The Jossey-Bass Handbook of Nonprofit Leadership and Management*, edited by Robert D. Herman and Associates, pp. 587–622. San Francisco: Jossey-Bass.

Mt. Auburn Associates. 2005. *Louisiana: Where Culture Means Business*. Baton Rouge: State of Louisiana, Office of the Lieutenant Governor, Louisiana Department of Culture, Recreation, and Tourism, Office of Cultural Development, Louisiana Division of the Arts, http://www.crt.state.la.us/arts/Publications/culturaleconomyreport.pdf (accessed November 23, 2011).

Ott, J. Steven. 2001. *Understanding Nonprofit Organizations: Governance, Leadership, and Management.* Boulder, CO: Westview Press.

Reid, Benjamin, Alexandra Albert, and Laurence Hopkins. 2010. *A Creative Block? The Future of the UK Creative Industries.* London: Work Foundation, http://www.theworkfoundation.com/assets/docs/publications/277_A%20creative%20block.pdf (accessed December 19, 2011).

Reider, Rob. 2001. *Improving the Economy, Efficiency, and Effectiveness of Not-for-Profits: Conducting Operational Reviews.* New York: John Wiley.

Roodhouse, Simon. 2006. "The Creative Industries: Definitions, Quantification and Practice." In *Cultural Industries: The British Experience in International Perspective*, edited by C. Eisenberg, R. Gerlach, and C. Handke, pp. 13–32. Berlin: Humboldt University Berlin, http://edoc.hu-berlin.de/conferences/culturalindustries/proc/culturalindustries.pdf (accessed December 19, 2011).

Rossiter, Nancy, Peter Goodrich, and John Shaw. 2011. Social Capital and Music Entrepreneurship. *Journal of Management and Marketing Research* 7: 1–12, http://search.proquest.com/docview/864625509?accountid=11441 (accessed November 15, 2011).

United Nations. 2010. "Creative Economy: A Feasible Development Option." *The Creative Economy Report*, http://www.unctad.org/en/docs/ditctab20103_en.pdf (accessed November 8, 2011).

Veblen, Kari K., and B. Olsson. 2002. "Community Music: Towards an International Perspective." In *The New Handbook of Research on Music Teaching and Learning*, edited by R. Colwell and C. Richardson, pp. 730–53. New York: Oxford University Press.

Wirtenberg, Jeana Abrams, and Carolyn Lilian Ott. 2004. "Assessing the Field of Organization Development." *Journal of Applied Behavioral Science* 40: 465–79.

Wirtenberg, Jeana, Thomas E. Backer, Wendy Chang, Tim Lannan, Beth Applegate, Malcolm Conway, Lilian Abrams, and Joan Slepian. 2007. "The Future of Organization Development in the Nonprofit Sector." *Organization Development Journal* 25, no. 4: 179–95, http://search.proquest.com/docview/197981069?accountid=11441 (accessed October 28, 2011).

Work Foundation. 2007. *Staying Ahead: The Economic Performance of the UK's Creative Industries.* London: Department for Culture, Media and Sport, http://www.theworkfoundation.com/DownloadPublication/Report/176_176_stayingahead.pdf (accessed December 16, 2011).

Chapter 11

The Birth and Development of a Cultural Advocacy Group in Hong Kong: Centre for Community Cultural Development

Chiu Yu Mok and Eric Ng

The term *community cultural development* (CCD) describes the work of artist-organizers and other community members collaborating to express identity, concerns, and aspirations through the arts and communications media. It is a process that simultaneously builds individual mastery and collective cultural capacity while contributing to positive social change. Other terms are used to describe such work. These include *community arts, community animation* (used in France), *community-based arts,* and *cultural work*. However, *community cultural development* encapsulates the salient characteristics of the work:

- *Community* acknowledges its participatory nature, which emphasizes collaboration between artists and other community members.

- *Cultural* indicates the generous concepts of culture (rather than, more narrowly, art) and the broad range of tools and forms in use in the field, from traditional visual and performing arts practice, to oral-history approaches usually associated with historical research and social studies, to the use of high-tech communications media, to elements of activism and community organizing typically seen as part of non-arts social change campaigns.

- *Development* suggests the dynamic nature of cultural action, with its ambitions of conscientization and empowerment, linking it to other enlightened community development practices, especially those incorporating self-development rather than development imposed from above.

To put it slightly differently, CCD is community art for community building through cultural activities. The cultural activities promote solidarity, dialogue, communication, tolerance, and diversity in cultural development in the face of the onslaught of the homogenizing force of globalization. The process and participation are meant to vest members of the community with the means of creative expression, enabling them to speak out.

CCD work turns the usual passivity of spectators in the arts into active producers of the arts. The workshop is the dynamic element in the process of community cultural development work, and the process is valued over the product. The workshop is a transformative forum through which individuals have the opportunity to change.

The difficulty in talking about community arts in China is that when it comes to the official level, you will be told that China is doing a lot of this already: "Look at the parks where people are waltzing and look at our children's palaces, where we have courses that pack them in. Arts by the disabled? Look at our Kuan Yin with *A Thousand Hands* performed by the deaf. You talk about the use of the arts to promote solidarity and communal feelings? Well, creating harmony is our very important goal." But community art in China today is not about empowerment, rights, social change, or even creativity.

If we might digress a little into the history of community arts in China, we can go back to one of the foremost theater artists/workers in China, Hung Fut Sai (the first chancellor of the Shanghai Theatre Academy). In the countryside of Hebeii Province, in the 1930s, Hung initiated a close encounter between Chinese peasants and modern theater. The theater experiment was part of a wider endeavor to promote literacy, economic development, health and sanitary improvements, and culture. Hung wrote plays about peasant conditions, at times fitting contemporary interpretations into old traditional stories, at times addressing the lives and concerns of the peasants, and at times even adapting Western works. At the beginning, Hung worked with the educated staff of the Organization Promoting Education for the Common People. Their performances were extremely popular, and the troupe was not able to meet all the performance demands. Then folks in one village formed their own theater troupe under the guidance of the organization and performed three plays in two nights. This lit the fire of enthusiasm in the peasants. Peasant theater groups spread like wildfire, and in 1933 and 1934, thirteen villages formally set up peasant theater groups, with more unofficial groups. Even more interesting was the fact that some peasants actually built a number of open-air theaters of varying designs. They also paved the way for subsequent experimentation in environmental theater and participatory theater, where actors would mingle with the audience and the audience/peasants would be able to intervene or participate in mass scenes.

In the early days of the People's Republic of China, there were high hopes and expectations. While cultural activities were not necessarily the first thing on people's minds, certainly the idea that every collective would enjoy a vigorous cultural life—with its own theater and music troupes, with its own Tian Han (a famous national playwright) and its own Nip Yi (a famous national composer)—captured the imagination of many.

According to *China's Feet Unbound* (1952), written by W. G. Burchett, an Australian journalist friendly to the Chinese Revolution, "things looked rather promising." He was referring to *Gate No. 6,* a play he saw in 1951 about Tietnsin dock workers and the smashing of the gang system. "It was necessary to show the public exactly how the system worked and how dockers lived," and "It took two nights to perform and was played by the dockers themselves." Burchett also described similar plays written and performed by ex-prostitutes and plays done by a former band of Tientsin pickpockets.

Writing more or less at the same time was Rewi Alley, a Kiwi admirer of the Chinese Revolution:

In Peking, the effect of bringing the people, old and young, literate and illiterate, into the theatre, has been a very marked one. ... The Chinese language is not easy to learn, and many people find it a formidable task. But the theatre does not allow thinking to lag behind. And as I sit and write this, from a compound nearby, I hear the orchestra of the railwaymen's theatre group practicing. The People's Liberation Army has its trained theatre group, for the Army today is very much a part of the life of the people. Schools and other organizations all have theirs. The theatre both educates and finds expression for all sections everywhere in China today, and belongs, in the truest sense, to the people themselves. Only a revolution could have produced the contrast between the theatre of today and the theatre of the past. [Alley 198]

Twenty years later, Lois Snow, the wife of a famous American sympathizer of the Chinese Revolution, wrote after her first trip to China:

"Theatre, music, song and dance remained an important part of people's lives. From school children to factory hands, from university students to farmer-peasants, every where we went some manifestation of this love of expression was obvious." Snow went on, in her book *China on Stage*, to describe middle-school kids putting on full-fledged shows of the Peking Opera and doing songs such as "Celebrate the Communique of the Second Plenary Session" and "People of the World Unite to Defeat US Aggression." Kindergarten children aged five and six dressed in replicas of PLA uniforms and caps and sang and danced in a program that included "Rely on the Masses" and "We Must Liberate Taiwan." A theater troupe of thirty-four hundred textile workers performed music and songs from the model operas *White Haired Girl* and *The Red Lantern* and Mao Zedong's poem "Snow." Shows ended with Little Red Books being waved while "Socialism Is Good" was sung. Unsophisticated farmers guided by two Peking students doing their manual labor in the countryside performed "I Just Heard on the Loudspeaker That the Soviet Revisionists Want Two Chinas!" Kids seven to fourteen years old with theatrical promise trained at a special school performed two other revolutionary operas: *Taking Tiger Mountain* and *Shachiapang*. Snow ended her book saying, "So it continued wherever we went; the successors to the revolution are full of song."

But unfortunately, as Lois Snow tells us, by the late 1960s the Chinese Revolution allowed people to sing only one song and see or perform only eight revolutionary operas.

After the Cultural Revolution, control over literature and art became more relaxed. With the opening up and gradual return of capitalism, full state subsidies for the arts and art troupes have gone away.

It is interesting to contrast what happened in the countryside of Hebei Province in the 1930s with the countryside in China today. In the countryside, we have seen no popularization of the theater or any art. Modern theater during the communist era since 1949 has not developed in the countryside, and traditional and folk theater and arts are being replaced by passive television and DVD watching and consumerist karaoke singing. In the cities, there are more choices—with movies and spectator arts being most common and readily available. And one wonders whether people in China today are culturally that much better off than in the days of the Great Proletarian Cultural Revolution of the 1960s and 1970s, when the whole nation was allowed to watch just eight revolutionary operas.

At one time, the national policy of the PRC on art and literature was based on Mao's promulgation at the Yenan Forum on Literature and Art in

1942: "The first problem is: literature and art for whom? All our literature and art are for the masses of the people, and in the first place for the workers, peasants and soldiers; they are created for the workers, peasants and soldiers and are for their use." And just recently, in May 2012, on the seventieth anniversary of the Yenan Forum, Hu Jintao, party secretary and state chairman, reiterated that art and culture are to serve the people and socialism, allowing a hundred flowers to blossom and a hundred thoughts to contend. Yet there was no mention of people's participation in the creation of the arts. As China is integrated into the global system, China seems to want to give the masses their bread and circuses—songs, dances, theater, and opera that they can see for free on TV. In the absence of democracy or any conscious effective advocacy for democratization of culture and the means of producing culture, the consumerist and market-driven logic of the "cultural industries" is heavily embedded in official thinking and practice. Investment in arts and culture is a cost that must bring profitable returns.

One wonders, then, how officials are responding to the Seoul Agenda, which was a major outcome of UNESCO's Second World Conference on Arts Education held in Seoul, Korea, from May 25 to 28, 2010.

The "Goals and the Action Plans for the Development of Arts Education" laid down in the Seoul Agenda are as follows:

1. Ensure that arts education is accessible as a fundamental and sustainable component of a high-quality education. Actions proposed include enactment of policies and deployment of resources to ensure sustainable and lifelong access to in-school and out-of-school experiences in all arts fields for a diversity of learners from different age groups and from all social backgrounds.
2. Ensure that arts education activities and programs are of a high quality in conception and delivery. Actions proposed include provision of sustainable professional learning mechanisms for teachers and artists; initiation of partnerships between artists and teachers in delivering curricula; and cooperation between community organizations and teachers in providing arts education programs in a variety of different learning environments, involving parents and family members. Community members and other stakeholders should rely on building partnerships to strengthen the role of arts education in society, especially across educational, cultural, social, health, industrial, and communication sectors.

3. Apply arts education principles and practices to contribute to resolving the social and cultural challenges facing today's world. Actions proposed include the recognition, fostering, and enhancement of knowledge and understanding of the social and cultural well-being dimensions of arts education through training programs, with emphasis on:

- the value of a full range of traditional and contemporary arts experiences

- the therapeutic and health dimensions of arts education

- the potential of arts education to develop and conserve identity and heritage as well as to promote diversity and dialogue among cultures

- the restorative dimensions of arts education in postconflict and postdisaster situations

This goal also includes fostering the capacity to respond to major global challenges on a wide range of contemporary social and cultural issues, such as the environment, global migration, sustainable development, and the fostering of democracy and peace to support reconstruction in postconflict societies through arts education.

We bring into the discussion the goals and action plans of the Seoul Agenda because while the document does not include the term *community cultural development* and while CCD work is not the same as arts education, the goals pronounced by the arts educators in Seoul are very similar to those that CCD workers pursue. CCD workers are arts educators, artists, organizers, and researchers. CCD activities—whether in the form of circle painting, community music, community theater (for example, playback and forum theaters), or dance—are processes that allow community members (however defined) to learn and eventually own the means of creative expression for aesthetic as well as social performances.

Dr. Sun Wai Chu, deputy head of the Shanghai Theatre Academy, was inspired by Richard Schectner, with whom performance studies originated. In performance studies, performance is treated as an object of inquiry and performance is seen as a lens through which to look at the world. Dr. Sun set up a Performance Studies Centre at the academy and promotes the concepts of aesthetic and social performance. The former refers generally to artistic activities that happens on stage—theater, dance, music, opera, puppetry, and so on, while social performance refers to the many activities undertaken by judges and lawyers in court, politicians and government officials, medical doctors, teachers, salesmen, reporters, and media personnel, or to interactions of any person in everyday life. Dr. Sun's program has produced graduates and postgraduates who are not just keen to be involved in the so-called theater district in Shanghai but who also have extensive training in social performance, including theater in enterprise, theater in education, theater in promoting mental health, and so on. Indeed the Shanghai Theatre Academy is doing a remarkable job, but the vastness of China posits the need for many other participants and similar training and services.

The Centre for Community Cultural Development, being a small organization, has focused on the following: theater, dance, music, visual arts, film, and arts-related therapies. (In due course we might expand into photography, video, and other art forms.)

We have been doing and providing training in community theater in many forms and different methodologies—for example, playback theater, forum theater, play forward theater, ethnodrama, and puppetry. Creating and devising their own plays, or using spectators as spec-actors, participants articulate their own stories.

We have been doing and promoting inclusive dance, which we call symbiotic dance. It was developed by local dancer Yuen Jie from a blending of the styles of DanceAbility (Alito Alessi), Transformance (Dan Baron Cohen), and contact improvisation. To the participants, rehearsals and practice bring the joy of dancing, the satisfaction of creating one's own movements, interaction with others, pride in one's own body, communication with the public, and new feelings related to the land. While performing, dancers celebrate their knowledge and interpret the histories of their own bodies, enthralling the imagination, expressing diversity, defeating prejudices, and helping build a more inclusive world and society. Everyone is dancing, united in collective performance and celebration, without many words. They are doing one dance but with many combinations and permutations. For many, dancing helps release tension and resistance, enabling people to speak out through a new kind of literacy. In the end, artists fully merge with the community, and one can discern the symbiotic growth of artists and communities.

Working with Peter Moser, a renowned British community musician, we utilize a simple and systematic approach with many diverse groups. Ordinary people create their own music and write their own songs as means of creative self-expression, often within a political and social context while working in an atmosphere of open collaboration, which lends itself to the forging of friendships, solidarity, and a sense of community.

Circle painting, an innovation by Vietnamese-American artist Hiep Nguyen, has been an extremely popular visual art activity in Hong Kong. The keywords for circle painting are *connect, create,* and *celebrate*. Circle painting brings people together to make art at social and community events. It provides an enlivening, engaging, and positive process for participants to make stunning large-scale paintings designed around the circle theme.

The first and second Hong Kong International Deaf Film Festivals, which included workshops and contests, were held in 2010 and 2012. The event allowed the people of Hong Kong to appreciate the achievements of deaf film artists, including those from Hong Kong. It allowed Hong Kong to become aware of the voice of the deaf through cinema. The event had good support from both the deaf community and the community at large. It brought increased awareness of the deaf, increased solidarity and pride among the deaf in Hong Kong, an enhanced identity of the deaf in Hong Kong, and the affirmation of deaf culture, of which deaf films are very much a part.We believe that proficient and effective CCD workers should be able to play the following roles: artist, teacher, organizer, and researcher. We provide training and opportunities to interested artists, social workers, and teachers who want to take part in growth-enhancing workshops in different kinds of arts-related therapies—music, drama, dance, expressive art, visual art, and so on—with local therapists (who are few in numbers and are all trained overseas). CCCD has also organized three major arts-related therapy conferences involving pioneering therapists from abroad giving keynote speeches and workshops. We have also tried to familiarize Hong Kong with the theory and practice of community cultural development in the art, social work, and educational communities and in society at large. We have trained trainers and then developed programs in Hong Kong to test out effective use of such methods and then to continue such activities as sustainable ones.

Without regular funding except from a limited number of donors (who authorize their banks to transfer money to CCCD on a monthly basis), we have been dependent on one-off project grants from the Hong Kong Arts Development Council of the Leisure and Cultural Services Department. Occasionally we tap some overseas funding through our overseas partners—for example, British Council funding through Peter Moser and his organization Moremusic.

Community cultural development work as pursued in Hong Kong does not have a firm foundation in Hong Kong. We have not found it easy to do CCD work in Hong Kong, although in the long run we believe that through our efforts and those of others, Hong Kong people will turn away from being only passive consumers of TV and pop culture and will learn that there is a better, more fulfilling alternative by becoming active creators of the arts themselves. Of course, we will need and hopefully soon get more artists, social workers, and educators with the skills that can lead beyond the "follow me" approach to allow the gold mine of creativity embedded in everyone to be truly revealed and unleashed. We will find and train more artists who are willing to work within communities and who are not geared toward the applause or glamour found on the stages of City Hall, the Cultural Centre, or the Hunghom Sports Stadium. We will persuade social workers, teachers, and their trainers that the arts are an additional means through which to serve and to empower. The people will find out how CCD works—that it is interactive, simple, and colorful. People will find that CCD activities are joyous community events that can happen at a school or at mass gatherings organized by social service groups.

Will CCD be meaningful and useful in Greater China? Being a small organization, we have not done a lot in China. But we have done workshops of various art forms (theater, dance, music) in different parts of China—mostly with NGOs, their staff, and their members. At other times, we have performed and exhibited.

Our staff and trainers (including artists from abroad) have worked in Beijing, Shanghai, Guangzhou, Shenzhen, Dongguan, Sichuan, and elsewhere. We have been able to do this through links to individuals, groups, and networks. For example, in 2007 the Congress of the International Drama/Theatre and Education Association invited representatives of groups and educational institutions to come to Hong Kong to attend workshops before and after the congress.

Almost all the workshops, mini-festivals, and trainings that we have attended or hosted happened on a rather ad hoc basis. We were invited to do something over a period of one to three days, although information about the activities was transmitted through networks instantaneously and people went Shenzhen, Guangzhou, or even Hong Kong for workshops, training, or activities.

We are not sure how to be more systematic or strategic in introducing CCD work in China. We are just beginning to think of ideas. Of course, there are the successful cases of the pop music, film, and TV industries, but they may not be exactly relevant. Perhaps we should look closer at the success stories of Willy Tsao (a Hong Kong dance enthusiast, founder of the City Contemporary Dance Company, who introduced modern dance to Beijing and Guangzhou), Edward Lam and his theater company, T. Z Chang and his visual art and film activities, and the educational efforts of Oxfam and World Vision. There have been failed enterprises: two well-known Hong Kong dramatists, James Mark and Dominic Cheung, moved north to set up Live Education with the Arts activities (with or without support of the children's palaces), but after a couple of years they retreated totally.

We believe that the opportunities are there and there are huge potentials for CCD for aesthetic and social purposes. Hong Kong has a number of advantages. There are CCD workers who can master the training and facilitation of CCD activities. Institutions like APA and the Hong Kong Art School are running relevant training courses for more CCD workers (although these courses need to be improved). Hong Kong is regarded as a leader in CCD work and has better connections with the rest of the world. Some private and public funding is available for promotion of CCD work in China. It is possible to tap Hong Kong government funding for small commercial enterprises that come under CEPA schemes. (Some playback theater workers have been able to get private funding for training trainers overseas.) It is time to explore further something that will change the cultural landscape of China.

BIBLIOGRAPHY

Alley, Rewi. *At 90: Memoirs of my China years*. Beijing: New World Press, 1986.

Burchett, Wilfred G. *China's Feet Unbound*. China: Lawrence & Wishart, 1952.

Snow, Lois Wheeler. *China on Stage: An American Actress in the People's Republic*. New York: Random House, 1972.

Chapter 12

Culture and the Media in a Global Age: The Role of the Media in Shaping Cultural Development in "Global City Wannabes"

Vivienne Chow

On October 25, 2013, more than one hundred thousand people rallied outside Hong Kong government headquarters at Tamar, Admiralty. It was the second protest in seven days drawing more than one hundred thousand participants. They were not shouting subversive slogans or demanding that certain officials step down. People took to the streets simply to ask the government why it had denied the right to more and better free-to-air television by rejecting a license application from a promising new TV station without a decent explanation.[1] But as this reality drama unfolded, the people of Hong Kong realized that what they had to fight for wasn't merely their right to television—they were defending the core values of Hong Kong and the city's cultural identity, which were not to be discarded despite Hong Kong officially becoming part of China after the handover of the city's sovereignty from Britain in 1997.

This free-TV license row leads to the question of where the media stands on city development—culturally, economically, and politically—particularly in Asia, where cities like Hong Kong, Singapore, Shanghai, Beijing, Taipei, and even Inchon in South Korea are up against each other in a global city race. What makes a city truly global without losing its local cultural characteristics? I would argue that media is the key.

This reality drama raised its curtains when three companies—Hong Kong Television Network (HKTV, formerly City Telecom), Hong Kong Television Entertainment (HKTVE), and Fantastic TV (FTV)—filed applications for domestic free-to-air TV licenses between December 2009 and March 2010, after the government decided to open Hong Kong's television market by removing a license cap in a television policy review in 1998.[2] For a long time, Hong Kong's TV market has been dominated by broadcasting giant Television Broadcasts (TVB) and its beleaguered rival Asia Television (ATV), which has been suffering from low ratings and hence low advertising revenue. (In 2011 TVB earned HK$2.8 billion and ATV made only HK$100 million.)[3] For most of the time, the free TV saga was an uneventful show, as the Executive Council—the Hong Kong Special Administrative Region's highest decision-making body, made up of the city's chief executive, top government officials, and society elites—could not make up its mind. This issue dragged on for two government terms: the granting of new free TV licenses was discussed when Donald Tsang Yam-kuen was chief executive; after his term was up and the new chief executive, Leung Chun-ying, came into office on July 1, 2012, there was still no news. While FTV and HKTVE had been lying low, HKTV, chaired by telecom maverick Ricky Wong Wai-kay, had been drumming up its presence, unveiling ambitious plans to produce TV dramas benchmarked against the standards of U.S. TV shows and investing a total of HK$900 million into production, talent, and operation. Wong's determination in realizing his TV dream drew wide support from the public, which had complained that TVB's domination had led to the homogeneity of Hong Kong's television culture and hegemony in Hong Kong's television market—TVB had 94 percent of Hong Kong's TV audience share according to its 2012 annual report.[4]

Just when it was least expected, the plot took a dramatic turn on October 15, 2013, when the Hong Kong government finally made up its mind. HKTV's application was denied without a clear explanation. Licenses were granted "approval-in-principle" to FTV and HKTVE, subsidiaries of i-Cable and PCCW, respectively. For their part, i-Cable and PCCW are existing pay-TV players owned by Hong Kong's most powerful "economic feudal lords"—i-Cable is a subsidiary of Wharf, while PCCW is owned by Richard Li, son of Hong Kong property tycoon Li Ka-shing, one of Asia's richest men.[5]

Then and there, the people of Hong Kong, who were notoriously apolitical after decades of training under British colonial rule, woke up as if they had been slapped on the face. They realized that their right to watch better TV—a humble wish for Hong Kongers, many of whom preferred good entertainment over politics—could be taken away by politics. The free-TV license applications first had to be approved by the Communications Authority (called the Broadcasting Authority before 2012), the statutory body that regulates broadcasting and communications sectors. The authority then submitted its recommendation to the Executive Council. If the Executive Council did not agree with the authority's recommendations or decided to change the policy from having no license cap to selecting two out of the three applicants, the whole licensing procedure would have to start all over again, with public notification of the change in policy. However, there was no such procedure. The government did not explain what HKTV did wrong, costing its bid. The Communications Authority later accused the government of ignoring its professional advice to issue licenses to all applicants.[6]

The public was outraged, criticizing the government for its lack of transparency. People were stunned by the fact that decisions could be made behind closed doors, with no public accountability. Subsequently, on October 20, 2013, reportedly 120,000 rallied against the government's unexplained decision. Demonstrators questioned why the government decided to go against the original TV policy stipulated in 1998, which promised to open the TV market without capping the number of licenses to be granted. They also questioned why HKTV, the most promising and daring in the production of high-quality entertainment, was ousted.[7] After losing the license bid, HKTV laid off 320 employees. Many of them camped outside central government offices for a week, with thousands of people joining the rally every night. Even TV and movie stars, including Chow Yun-fat and Andy Lau, who had been notoriously shy about making political comments, stood up to voice their discontent. It was the peak of this long-running reality series.

The matter remained unresolved by mid-November 2013, despite repeated rallies. Lawmakers' motions to probe into the Executive Council's decision were voted down by pro-Beijing legislators, who hold the majority of seats in the Legislative Council, Hong Kong's lawmaking body. Nevertheless, this incident can well be recognized as one of the largest political events originated from a cultural cause, with people using political means (rallies, protests, and debates in the Legislative Council) to achieve a cultural end (better television and an end to TVB's cultural domination). People care because TV is more than just entertainment. TV—like the press, film, music, and design—is a cultural product and arguably the most powerful media to impact our daily lives. Cultural products take up our time when we are not sleeping or working. They tell stories that reflect our way of life—a sense of culture as defined by Raymond Williams.[8] They shape our perspectives of the world and attitudes toward the society we live in. They play "a key role in changing our individual awareness, cultural attitudes, and even public policy … [putting] events in context, helping us to better understand both our daily lives and the larger world."[9] These stories construct a common meaning through "symbols, images, values, ideas and beliefs" shared by viewers, who are mostly ordinary people, and such common meanings will therefore gradually form a society's cultural identity—where we come from, where we belong, and how we identify ourselves.[10]

Hong Kongers identify themselves with Hong Kong cultural products reflecting their way of life and their values. Hong Kong's TV culture peaked in the 1970s. Creative shows, including dramatic series and even experimental works (such as Patrick Tam's *Seven Women* series on TVB, which paid homage to French New Wave director Jean-Luc Godard) produced by TVB, ATV, and the now defunct Commercial Television, reflected Hong Kong's economic boom and subsequent cultural implications. Hong Kong films, from Jackie Chan's kung fu comedies to John Woo's highly stylized action thrillers from the heyday of the 1980s, defined a new film genre and depicted the creativity and lively energy of Hong Kong—called one of the four Asian "little dragons," alongside Singapore, South Korea, and Taiwan. These TV shows and films—featuring Canto-pop—were exported to Southeast Asia, Taiwan, Korea, and Chinese-speaking communities around the world. Hong Kong was Asia's cultural capital along with Japan, with diversified cultural exports. Hong Kong pop culture put Hong Kong on the culture map. Hong Kong pop culture was Hong Kong's pride, just like the Korean Wave is to Koreans today.

Justin Lewis, professor of communications at Cardiff University, argues that a free market for cultural industries is a failure because the free market is not entirely free. It is subject to market censorship under the force of advertisers, which pay the bill for the creation of cultural products.[11] In the United Kingdom, there is the British Broadcasting Corporation (BBC), which is advertisement-free and survives on license fees paid by the public. But Hong Kong doesn't have such a publicly funded station that runs two full channels. (The publicly funded Radio Television Hong Kong produces only programs carried by TVB and ATV. The station plans to launch three digital terrestrial TV channels in 2014. TVB continued to grow thanks to economies of scale, but ATV stopped producing dramas in the first decades of the 2000s because of low advertising revenues. This means that in Hong Kong, if you can't afford the 360 (mostly foreign) channels offered by pay-TV stations, you don't have that many options for local free television. Thus TVB became the dominant player in the city's television industry, and mainstream television became a monocultural sphere. People have grown tired of the cultural stagnation created by TVB's near monopoly. Instead of promoting new ideas and cultural diversity, the station produces dumbed-down TV shows aimed at the largest mass audience to suck up to advertisers (with excessive product placement, for example). "TVB reality" and "TVB values" are detached from real life. For instance, many dramas revolve around family disputes over a massive inheritance from a rich man between the families of his two wives, even though the concubine law was abolished in Hong Kong in 1971. On TVB shows, people doing normal office jobs live in massive flats, despite the fact that property prices in Hong Kong have skyrocketed, and the game show *Super Trio* has gone on for eighteen years.[12] Statistics show that Hong Kong television viewers are gradually leaving TVB. In 2005 the Korean period drama *Jewel in the Palace* set the record for Hong Kong TV ratings when its finale was aired on TVB, peaking at a top-minute rating of 47 TVRs (television viewer ratings)—equal to 3,055,000 viewers, almost half of Hong Kong's 7 million people at the time. The show as a whole had an average rating of more than 35 TVRs. *Jewel in the Palace*'s peak rating broke the record set by the 2000 TVB sitcom *War of the Genders*, which attracted 3,018,000 viewers at its peak minute. However, eight years later, in 2013, only two TVB dramas had more than 30 rating points (or 1.95 million viewers) on average. The average rating that year went down to about 25.6 (or 1.6 million viewers). Where have the viewers gone? In my case, I have chosen to watch American or British dramas on the Internet, but I can't find many friends to share the experience with, as we all watch different shows. Do I feel disconnected with Hong Kong? Yes. But why should I watch shows that I don't enjoy? Still, the government refuses to introduce a player that will produce shows to not only bond Hong Kongers again but also to challenge the status quo maintained by TVB. How can people ignore this?[13]

More importantly, this TV saga has revealed how the governance of Hong Kong has changed since the handover of the city's sovereignty from Britain to China: a government that has Beijing pulling the strings behind is crushing a governance system inherited from British rule by arbitrarily rejecting professional advice conducted by a statutory body in its decision-making process. This poses a serious threat to the core values of Hong Kong—the rule of law and freedom of expression, which do not exist in mainland China. The core values are intrinsic elements in cultural representation of a society reflected in the media and many other cultural products. Media is culture and a cradle for ideas and creativity. The majority of people who rallied were defending Hong Kong's culture and cultural identity—the cultural end of this TV saga.

A city that brands itself "Asia's world city,"[14] Hong Kong needs a healthy cultural ecology driven by a diverse and creative media scene that is protected by the right to freedom of expression. If television is the most accessible medium and is an important component of a local culture, if watching television is what most people do, then what kind of television—or culture in a broader sense—do we want for the city? Lewis argues that "a realistic cultural policy ... means engaging with the things that our society does, not with what we feel our society ought to be doing. Most people watch TV; they do not, as a rule, go to watch plays. A cultural policy should begin by devising ways to improve television rather than dreaming up fruitless plans to drag people back into the theatre."[15] And what kind of role should the media—serving as a mediator between policy makers, content producers, and the public—play in fostering the growth of a city's culture?[16]

As we have moved into the global age of the twenty-first century, cities, as opposed to nations, have emerged to play a much more significant role in a globalized world. This is different from the mid-1900s Keynesian period, when cities were merely "centres for administration, small-scale manufacturing and commerce."[17] But since the 1980s, cities have evolved from clusters of economic and financial activity bringing in a concentration of wealth. Some cities have risen to become global cities, attracting not only investment from around the world but also talent—what author Richard Florida calls the creative class—seeking new and exciting opportunities and a better lifestyle. Florida argues that "cities are cauldrons of creativity," which is a key driver of economic growth and development of a society. This creative class is a highly mobile class, typically made of globe-trotters. They will settle only in cities that not only pay a good salary but also offer an environment that is innovative, diverse, and tolerant.[18]

In turn, cities transform themselves to attract talents that will power their economic development. Saskia Sassen sees this as a rebuilding of cities into "platforms for a rapidly growing range of globalised activities and flows, from the economic to the cultural and political," where architecture, urban planning, and urban design take on major roles.[19]

Under the force of globalization, global firms are planting their flags across the world. We see the same corporations and chains operating in different cities on different continents. If globalization is a long-running series, then these firms are the stars of the growing shows. They need more and more platforms to showcase their talents, creating an increasing demand for global cities.[20]

Many cities are happy to take part in this game. From Birmingham to Amsterdam and to Asian cities like Hong Kong, Shanghai, and Singapore, cities are eager to hop onto this global circuit. Besides attracting international investments, cities seek global cultural icons and institutions. We see buildings by the same "starchitects"—Zaha Hadid, Frank Gehry, Norman Foster, and Herzog and de Meuron—in many of these "global city wannabes." Take Hadid, for example. The Iraqi-British architect has been leaving footprints across Asia over the past decade, and her name is associated with new cultural landmarks: Innovation Tower in Hong Kong, D'Leedon in Singapore, Guangzhou Opera House, and the Abu Dhabi Performing Arts Centre. I can't help but wonder: is Zaha Hadid the new Starbucks? The same "global elites"—from the finance industries to cultural experts—have become the new globe-trotting species, as they can get the same top jobs anywhere. Examples include Michael Lynch, former chief of London's Southbank Centre, who became the chief executive officer of West Kowloon Cultural District in Hong Kong. Global art fairs such as Art Basel are making new destinations in new cities; MCH Swiss Exhibition (Basel), organizer of Art Basel, the world's largest modern and contemporary art fair, acquired a 60 percent stake of ART HK, organizers of Asia Art Fairs, in 2011. Art Basel expanded to America in 2002 with the first Art Basel Miami Beach. In May 2013 Art Basel was inaugurated in Hong Kong with the sponsorship of Deutsche Bank. Toward the end of the Hong Kong fair, organizers announced that the Swiss bank UBS, which had sponsored Art Basel Switzerland and Art Basel Miami Beach, would replace Deutsche Bank to sponsor the Hong Kong fair, completing the globalization of the Swiss art business.[21]

Cities that are not graced by the presence of these global brands are actively following the same model to create their own versions of Art Basel: Art Stage in Singapore, Art Taipei, Art Beijing, Art Tokyo, and so on. Even cultural institutions are actively reaching out through collaborations in cultural projects or are setting up new branches to "conquer" a new frontier in Asia (for example, a new Guggenheim Museum in Abu Dhabi).

Asian cities strive to achieve global city status by investing heavily in the arts and cultural infrastructure (museums, concert halls, and even cultural districts) from a top-down model: in Hong Kong there is the HK$21.6 billion (US$2.8 billion) West Kowloon Cultural District, where many new performing arts centers and a new visual culture museum called M+ will be built. The first phase is expected to open in 2017. Heritage sites such as the Central Police Station and the PMQ (former Police Married Quarters) will open in the next few years as new cultural and creative destinations in Hong Kong (yet the government hasn't even bothered to fix Hong Kong's TV culture, as I have already explained). In Singapore there is the new National Art Gallery opening in 2015, as well as the Gillman Barracks, where galleries and the new Centre for Contemporary Art are located. That's in addition to many private museums built at light speed all over China.

The opposite of global is local. To make space for globalization, certain local things have to be sacrificed. This leads to gentrification, urban redevelopment, and so on. Old districts are replaced by shiny buildings, which are signs of modernity. Local indigenous cultures are at stake. They are slowly being eradicated under the force of globalization and the aspirations of those grand schemes and structures.

The new wave of cultural globalization and vast cultural development has hit Hong Kong hard. Yet the local media has yet to figure out how to react. Government statistics show that by the end of July 2013, Hong Kong had fifty-three daily newspapers and 701 periodicals, including electronic newspapers, two domestic free-to-air TV stations (the two new free-TV stations had their licenses granted only in principle), three domestic pay-TV stations, eighteen nondomestic TV stations, four sound broadcasting licensees, and Radio Television Hong Kong, the government-funded public-service broadcaster.[22] Of these media, few are specialized in culture. Most of them are mass media covering general news from a traditional hard-news versus soft-news perspective: politics, crime, the environment, welfare, education, health, labor, and land development are hard news and the main beats in a traditional news environment, and these stories go to the front book, the most precious spot in a traditional newspaper. Culture stories, on the other hand, are considered soft and are often pushed to the periphery, to the lifestyle section, covered from a lifestyle angle. These include profiles of celebrities and artists and event previews that serve as activity guides. Such a content alignment sends audiences a message that culture has a much lower priority in the media and hence low priority in society. Most frustrating in my personal experience as a journalist is that those who cover hard news are not bothered to take culture seriously, while those who write about culture are in fact content with the current setting and have no interest in bringing a cultural perspective to the hard news section. This is why coverage of the West Kowloon Cultural District is handled by the political team in some media organizations, since disputes over the spending of public money sound sexier to journalists. They fail to juxtapose such public spending on culture against a broader picture of Hong Kong's cultural ecology. Competition over advertising dollars also leads to the decline of art criticism in the mainstream media, as the areas it covers—visual arts, performing arts, and literary arts—are deemed "high culture" and "elitist" and often rate much lower than sexy crime stories. On TV, these are the topics that get the worst time slots. In print, these are the items that land on the worst possible pages, if they are lucky. If not, they are nonexistent in today's media. Advertisers want to place ads on pages that draw eyeballs. Stories that no one reads make a page redundant from a business perspective. Why bother printing a page that doesn't make money?

Culture, however, is a resource—"raw material of the city"—and creativity is "the method of exploiting these resources and helping them grow" writes creative city theorist Charles Landry. "Culture, therefore, should shape the technicalities of urban planners rather than be seen as a marginal add-on to be considered once the important planning questions ... by contrasting a culturally informed perspective should condition how planning as well as economic development or social affairs should be addressed."[23]

Hence, to safeguard our cultures and our individual cultural identity, the media has to take up the role of education and must monitor governmental cultural policy development and implementation. The media plays an important role in fostering cultural development and inspiring creativity among citizens—constructing a unique city identity that allows one city to stand out from the rest. The media is not only part of culture; it is a form of "cultural practice" as defined by Williams in 1965.

Amid this new wave of cultural globalization and extensive cultural development, how do we convince media management that the traditional approach to news gathering and beats can no longer reflect reality?

Andrea Zlatar, founder of the cultural magazine *Zarez,* published in Croatia, has made a few suggestions for the role of the media in today's cultural ecology. She argues that the media, being a form of cultural practice, should be more proactive when it comes to culture. In her view, the media is an instrument of cultural policy, "a mediator between ... culture and society, ... conveying cultural products from the producers to the consumers."[24] To be precise, the media is a mediator between policy makers (cultural policy makers), cultural producers, artists, thinkers, and a wider public, conveying important cultural messages that will inspire their creativity. The media plays an indispensable role in today's cultural ecology, and this role must be recognized and taken seriously.

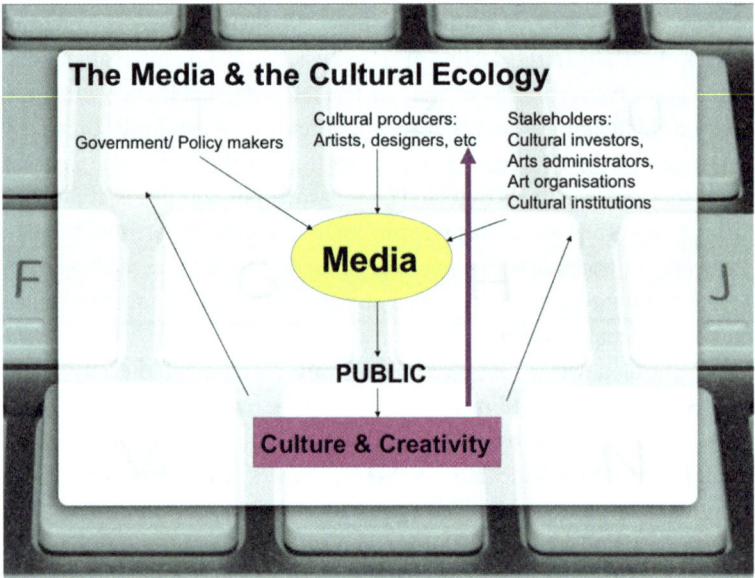

Figure 12.1 The Media & Cultural Ecology

But how can we achieve this? Can we look at culture differently in the media? Why can't a major exhibition opening in the city be reported as a news item, for the works of art will serve as an inspiration for many? Why do we report on the number of visitors and the astronomical cost of insuring some Pablo Picasso paintings exhibited in Hong Kong but not explain in the news why these paintings are artistically, historically, and culturally important? Can we have reporters who understand culture cover cultural policies instead of leaving them in the hands of political reporters, so that public spending on culture can be reviewed culturally rather than via accounting principles? Is it possible to cover the free-TV license row from a cultural perspective instead of merely a political perspective? Culture should be defined in its broadest sense among the media. It is not high culture but people's way of life. And to bridge the gap between people at all levels, the media must remain open and liberal so as to take different views and perspectives on reporting.

Maintaining a free and open media environment that promotes opinion and cultural diversity is fundamental, because only freedom of expression enables culture and the arts to live. However, although many Asian cities are actively investing in cultural development, they still hold a tight grip on the media. According to Reporters Without Borders' "World Press Freedom Index 2014," press freedom in Asia is notoriously low (as usual). Only four countries or territories (mostly home to so-called global cities) made it to the top 25 percent: Taiwan ranks at number 47, South Korea at 50, Japan at 53, and Hong Kong—sixteen years after the handover from Britain to China—down four points to number 58 (Figure 12.1). Authoritarian states all occupy positions 100 and above. Ironically, these are the states (including China at 173, the seventh lowest in the list) that are most aggressive in developing global cities, through various national policies and investments in cultural and creative industries, to construct their "soft power," eyeing the economic benefits this can potentially generate.[25] But the truth is that cities where press freedom is under threat will never succeed in transforming themselves into creative cities, which require a great deal of freedom to generate the creativity that feeds its own functioning and further growth.

To foster cultural development and inspire creativity, the media must remain independent. Not only must it refrain from favoritism (whether toward policy makers or commercial entities), but it must also promote a diversity of public opinions. Only the free flow of information and sensible reasoning in the public discourse can promote the culture and creativity to be applied to various areas of a city's development. Freedom of speech and expression are also key to any cultural and artistic development, as cultural and artistic creations are often critiques of a society's status quo.

To foster cultural development in a global age, developing a creative city and maintaining a healthy media industry are key. The media, whether general news outlets or agencies dedicated to cultural journalism and arts and cultural criticism, should be guaranteed press freedom and editorial independence.

We should embrace a pluralistic urban cultural identity as opposed to a dated national cultural identity. This vibrant, pluralistic urban cultural identity—built upon rich cultural resources, the arts, architecture, design, and, most importantly, a sophisticated media industry that promotes freedom of speech, diversity of opinion, and the free flow of information—in turn will attract more talent and their associated wealth and experiences to the city. The city will be the place where people want to live. We should aim at building a unique creative city heralded by a unique cultural identity instead of surrendering ourselves to globalization.

NOTES

1. "Hundreds of thousands of citizens plan to gather at the government headquarters and surround the Legislative Council" "萬計市民再聚政總劉德華林夕撐港視工會今撤離下月圍立會," *Ming Pao Daily News*, October 26, 2013, A1.

2. "Two TV licences got not but one denied," *South China Morning Post*, October 16, 2013, A1.

3. Amy Nip and Vivienne Chow, "Titan defends the turf," *South China Morning Post*, November 26, 2012, A4.

4. Vivienne Chow, "Calling 'cut' on a television dream," *South China Morning Post*, October 17, 2013, A4.

5. Alice Poon, *Land and the Ruling Class in Hong Kong* (Richmond, BC: Alice Poon, 2005), 10–15. The chief executive in council granted "approval-in-principle" to Fantastic TV and Hong Kong Television Entertainment, meaning that licenses were not officially granted to the two television stations. Further reviews of the two applicants are needed to decide whether the licenses will be officially granted.

6. Vivienne Chow and Tanna Chong, "Government 'ignored' TV advice," *South China Morning Post,* November 9, 2013, A1.

7. "This is not just about television," *South China Morning Post*, October 21, 2013, C1. Chan Pui Man, "Twelve thousand people shout loudly that they wish to have choices. 陳沛敏. "12萬人高呼：我要有得揀" *Apple Daily*, October 21, 2013, S09.

8. Raymond Williams, "Culture Is Ordinary," In *The Politics of Culture*, ed. Gigi Bradford, Michael Gary, and Glenn Wallach (New York: New Press, 2000), 17.

9. Richard Campbell, Christopher R. Martin, and Bettina Fabos, *Media and Culture* (Boston: Bedford/St. Martin's, 2012), 13.

10. Richard Kurin, "The New Study and Curation of Culture," In *The Politics of Culture,* ed. Gigi Bradford, Michael Gary, and Glenn Wallach (New York: New Press, 2000), 339.

11. Justin Lewis, "Designing a Cultural Policy," *Journal of Arts Management, Law and Society* (Summer 1994): 41–56.

12. Vivienne Chow, "Triumph in the Skies, Failure in Hong Kong," *South China Morning Post,* August 20, 2013, http://www.scmp.com/comment/blogs/article/1298068/triumph-skies-failure-hong-kong

(accessed April 18, 2014). "Television Broadcasting in Hong Kong," Commerce and Economic Development Bureau, http://www.cedb.gov.hk/ctb/eng/broad/tv.htm (accessed November 12, 2013).

13. Vivienne Chow, "Finale puts biggest jewel in broadcaster's crown," *South China Morning Post*, May 4, 2005, C3.

14. Brand Hong Kong, http://www.brandhk.gov.hk/en/#/en/about/overview.html (accessed November 12, 2013).

15. Lewis, "Designing a Cultural Policy."

16. Andrea Zlatar, "The role of the media as an instrument of cultural policy, an inter-level facilitator and image promoter: Mapping out key issues to be addressed in South East Europe," *Policies for Culture* (2003): 1–5.

17. Saskia Sassen, "Cities in Today's Global Age," *SAIS Review* 29, no. 1 (Winter–Spring 2009): 3

18. Richard Florida, *Cities and the Creative Class* (New York: Routledge, 2005), 1–34.

19. Sassen. "Cities in Today's Global Age," 5.

20. Ibid., 7.

21. Vivienne Chow, "Swiss bank UBS sponsors Art Basel Hong Kong," *South China Morning Post,* May 27, 2013, B4.

22. "The Media of Hong Kong," *Hong Kong: The Facts,* http://www.gov.hk/en/about/abouthk/factsheets/docs/media.pdf (accessed November 13, 2013).

23. Charles Landry, *The Creative City: a Toolkit for Urban Innovators,* 2nd ed. (Sterling, VA: Earthscan, 2008), 173–75.

24. Zlatar, "The role of the media."

25. "2014 World Press Freedom Index," Reporteres Without Borders, http://rsf.org/index2014/en-index2014.php (accessed 14 July 2014).

Chapter 13

The Development of Creativity in Hong Kong: Technology, Talent, and Tolerance

Victor Kwong

Richard Florida identified a new socioeconomic class, the creative class, in his book *The Rise of the Creative Class,* published in 2002. The steady economic shift of society in the developed world gave rise to this class because this group perceived that its work productivity could be meaningful and bring economic growth. Unlike traditional categories of social classes, the creative class was defined by its introducer. The creative class consists of two main divisions of professionals: creative professionals, who have advanced knowledge and are skillful in a particular career, and the supercreative core, a group of people dedicated to creative processes, such as investigations of new products.

The author also introduced the measurement and assessment of creativity's contribution to the regional economy via the Creativity Index.[1] Three components dominate the index: talent, technology, and tolerance, known as the 3Ts. The author discusses the relationship between the 3Ts and the Creativity Index in terms of the contribution of creativity to economic development in the past decade.

The purpose of this chapter is to investigate creativity in Hong Kong in the past decade by evaluating each factor of the 3Ts in Hong Kong. The paper also discusses the improvement of the current environment and society to retain and draw the creative class to Hong Kong.

THE 3TS THEORY OF ECONOMIC DEVELOPMENT AND THE CREATIVITY INDEX

In *The Rise of the Creative Class*, Florida writes that the 3Ts (talent, technology, and tolerance) are keys to understanding how creativity contributes to economic development in a knowledge-based working environment.[2] Unfortunately, these factors alone are insufficient to support an assessment of creativity. The following section introduces the three factors using the definition in the report *Creativity and Prosperity: The Global Creativity Index* published by the Martin Prosperity Institute of the University of Toronto.[3]

Florida clearly illustrates the meanings and implications of each T as well as several variables from the Gallup Organization's World Poll.[4]

Technology

Technology includes the following aspects:

- Research and development (R&D) investment: by the measurement of R&D spending as a share of GDP

- Research: by the measurement of professional researchers' level of engagement in R&D per million capita

- Innovation: by the measurement of patents granted per capita

Talent

Talent embraces the following dimensions:

- Human capital: by the measurement of higher formal educational attainment in the form of enrollment in postsecondary education

- Creative class: by counting the share of the labor force that engages in a high degree of problem solving in everyday work, including those working in computer science, mathematics, architecture, engineering, physical science, social science, education, training, library science, art and design, entertainment, sports, media, management, business, finance, law, sales, and health care

Tolerance

Tolerance is composed of the following two dimensions:

- Tolerance toward ethnic and racial minorities, as determined by the question, "Is your city or area a good or bad place for ethnic and racial minorities?"

- Tolerance toward same-sex relationships, as determined by the question, "Is your city or area a good or bad place for gay and lesbian populations?"

Created from the findings of *Creativity and Prosperity,* the Global Creativity Index is closely correlated with GDP per capita.[5] Moreover, talent (0.78 of the GDP per capita) has the most significant correlation to GDP per capita, followed by technology (0.72) and tolerance (0.63).

CREATIVITY IN HONG KONG FROM THE PERSPECTIVE OF THE 3TS

The following section investigates creativity in Hong Kong in the past ten years by evaluating each factor of the 3Ts. Talent is the most significant factor and is therefore measured using data about educational attainment (highest level completed) and occupation from the 2001 population census, the 2006 by-census, and the 2011 census.

Talent

Table 13.1 shows percentages of (1) the creative class population with postsecondary qualifications among the entire working population with postsecondary qualifications; (2) the creative class population with postsecondary qualifications among the whole creative class population; and (3) the creative class population among the whole working population in 2001,[6] 2006,[7] and 2011.[8]

Table 13.1 Comparison of the creative class population and the whole working population

	2001	2006	2011
Percentage of the creative class population with postsecondary qualifications among whole working population with postsecondary qualifications	78.5%	70.9%	71.5%
Percentage of the creative class population with postsecondary qualifications among the whole creative class population	54.5%	57.5%	69.9%
Percentage of the creative class population among the whole working population	31.6%	33.0%	35.4%

Source: Population censuses (2001, 2006, and 2011)

Florida highlights the common mistake of counting the working population with high educational attainment by counting the population of the creative class.[9] He says that 72 percent of the working population with college degrees are members of the creative class but less than 60 percent of the U.S. creative class has a college degree (that is, four out of ten members do not obtain college degrees). However, the percentage of the creative class with postsecondary qualifications among the whole creative class population has increased from 54.5 percent in 2001 to 57.5 percent in 2006 to 69.9 percent in 2011, which strongly shows the trend of more members of the creative class obtaining postsecondary qualifications in Hong Kong in the recent decade, in contrast to the trend in the United States. The percentage of the creative class with postsecondary qualifications among the whole creative class also tells us that members of the creative class who did not obtain postsecondary qualifications have participated in further postsecondary studies or even higher levels of study.

Moreover, Florida mentions that the creative class is highly mobile, moving to different places in the United States, not only for job opportunities but also for openness and diversity.[10] In Hong Kong the percentage of the creative class among the whole working population has increased steadily, from 31.6 percent in 2001 to 33 percent in 2006 to 35.4 percent in 2011. That is, the net movement of the creative class is positive, since the size of the whole working population also increased during those ten years. This reflects the idea that members of the creative class might choose to remain in Hong Kong because of the open and diverse environment.

The percentage of the creative class with postsecondary qualifications among the whole working population with postsecondary qualifications dropped from 78.5 percent in 2001 to 70.9 percent in 2006 and then slightly increased to 71.5 percent in 2011. This might indicate the expansion of postsecondary education, especially the exponential growth of subdegree programs from 2000 to 2006, providing more opportunities for the noncreative class to obtain higher education qualifications and then to change careers. An increase in occupations categorized as the creative class, as a percentage of the whole working population, might also explain this upward mobility of the original noncreative class.

Apart from using census data to investigate the creativity of Hong Kong in the category of talent, *Creativity and Prosperity* also posits the level of talent in Hong Kong.[12] Among the eighty-two nations assessed by the index, Hong Kong's global talent ranking was at 37, and it comprised the variables of human capital (ranked at 50) and the creative class (ranked at 26).

Technology

According to *Creativity and Prosperity*, only seventy-five nations were ranked under the factor of technology, and Hong Kong was ranked 22. The ranking was composed of the variables R&D investment (ranked at 41), researchers (ranked at 26), and the creative class (ranked at 10).

Tolerance

From the perspective of tolerance on the Global Creativity Index, Hong Kong was ranked at 12. The ranking included the variables of tolerance toward racial and ethnic minorities (ranked at 11) and tolerance toward gays and lesbians (ranked at 26). Hong Kong was ranked number 1 among neighboring nations in East Asia and Southeast Asia.

IMPROVEMENT OF THE ENVIRONMENT FOR RETAINING THE CREATIVE CLASS IN HONG KONG

Besides the 3Ts ranking, Hong Kong was ranked number 20 among the eighty-two nations on the overall Global Creativity Index. Although Hong Kong's ranking was relatively high, there is room for improvement for the sake of retaining the creative class, particularly in certain areas.

On the Global Creativity Index, in the area of technology, Hong Kong is ranked lower than Singapore by one, but Hong Kong is still relatively innovative because of R&D investment and the population of researchers (see Table 13.2). If more resources were allocated by the Hong Kong government with regard to R&D investment and researchers in both private and higher educational institutions, the overall technology ranking would move upward.

Table 13.2 Global technology rankings in Singapore and Hong Kong

	R&D Investment Ranking	Researcher Ranking	Innovation Ranking	Overall Technology Ranking
Singapore	11	4	11	10
Hong Kong	41	26	10	22

Source: Florida et al., *Creativity and Prosperity*.

From the perspective of the talent factor, this index measures (1) the size of the population with higher educational attainment at the postsecondary level (the variable of human capital) and (2) the size of the creative class population (the variable of the creative class). One limitation of the index is that it fails to measure the level of creative ability. The findings of this index show the people of Hong Kong lacking in creativity. A convincing argument among Hong Kong students is that they are not creative at all. This issue has been discussed for a long time. Needham argues that the scientific spirit is inborn but that the scientific method for exploring natural science is missing in traditional Chinese society.[13] Traditional Chinese culture, together with schooling, might limit children's imaginations, curiosity, and explorations.

Schooling

The variable of human capital is used to measure educational attainment but not to measure creative ability. Educational systems in the eighty-two nations included in the Global Creativity Index are different and lead to various levels of creative ability, even though citizens in different countries may attain the same level of education. Therefore, the ranking of human capital might show findings opposite the general comment on creativity in Hong Kong, especially since local students have been criticized by the mass media and employers for being narrow-minded due to rote learning approaches in the entire educational system in Hong Kong.[14] Therefore, it is essential to develop and establish new and better pedagogies in order to nurture students with both robust knowledge and creativity. In fact, some secondary and primary schools have reformed their curricula to enhance the creativity of students. But more work will be needed. Moreover, the exam-oriented or GPA-score-oriented culture in postsecondary education constrains the development of creativity as measured by the variable of human capital.[15] Higher education institutions should also take action to rectify this situation.

Parenting

Besides schooling, appropriate parenting—including nurturing creative ability—is seen as indispensable to the whole-person development of children. Inappropriate parenting causes barriers to developing children's creativity. This phenomenon is common in Hong Kong.[16] By sending children to extracurricular activities in kindergarten, like playing musical instruments, painting and drawing, and public speaking, Hong Kong parents love to build up their children's portfolios with bonus points for admittance to preferred primary schools. These skills are not viewed as academic requirements for admission to primary schools; instead they constitute "the school's potential source of pride" through winning trophies in ballet, swimming, and mathematics contests. Finally, children may lose their curiosity for exploration and opportunity for creative development. Although primary schools may take the blame for discouraging creative development of children, parents are actually responsible for providing the appropriate environment for the whole-person development of their children, including the development of creativity from early childhood.

FURTHER STUDIES ON THE DEVELOPMENT OF CREATIVITY IN HONG KONG

This chapter has adopted the 3Ts theory of economic development to investigate the question of creativity in Hong Kong in the past decade. However, several interesting areas might also affect relationships between creativity and Hong Kong's working environment: (1) facilitating growth of the ratio of the creative class to the working population and attracting the creative class to Hong Kong; (2) correlations between growth and decline of the population among different age groups and the creative class; and (3) the tensions and dilemmas between teaching knowledge and creativity. In fact, since the early 2000s the Hong Kong government has allowed mainland talents to emigrate to and reside in the HKSAR. This policy aims at directly or indirectly strengthening the size of the creative class in Hong Kong.

This chapter has discussed the strong relationships between creativity and talent, while the other 2Ts have strong connections with the Creativity Index. The data were drawn from various sources, and there have been no detailed studies in the past. A pilot study of these 2Ts in the context of Hong Kong has high potential and it is strongly recommended for the near future.

CONCLUSION

Although Hong Kong is not ranked in the top ten in the Global Creativity Index, it has been ranked third among Asian nations in the overall ranking. By using the 3Ts, this chapter provides a clear picture of creativity in Hong Kong, which should not be viewed negatively as lacking creativity. Hong Kong is not a dry desert without creativity. I suggest enhancements regarding creativity in Hong Kong, including (1) heavier investment in R&D and researchers for further development and improved performance in technology and its ranking; (2) the development of new pedagogies to nurture students' knowledge and creativity; (3) the reversal of exam-oriented learning in postsecondary education; and (4) the promotion of appropriate parenting styles. If these measures were really taken, the government and its citizens would benefit from the outcome and outputs of Hong Kong as a whole.

NOTES

1. See Richard Florida, *The Rise of the Creative Class: And How It's Transforming Work, Leisure, Community and Everyday Life* (New York: Basic Books, 2002).

2. Ibid.

3. See Richard Florida, C. Mellendar, K. Stolarick, K. Silk, Z. Matheson, and M. Hopgood, *Creativity and Prosperity: The Global Creativity Index*, 2011, http://martinprosperity.org/media/GCI percent20Report percent20Sep percent202011.pdf (accessed December 13, 2013).

4. See "Gallup Organization's World Poll, 2010," *Gallup Organization*, http://eu.gallup.com/poll/118471/world-poll.aspx, cited in Florida et al., *Creativity and Prosperity*.

5. Florida et al., *Creativity and Prosperity*.

6. "Working population x Educational attainment (highest level completed) x Occupation [Sub-major group]," *2001 Population Census*, http://www.censtatd.gov.hk/freedownload.jsp?file=stat_table/population/D5320039E.csv&title=Working+population+by+Education+attainment+(highest+class percent2flevel+completed)+and+Occupation+ percent5bSub-major+group percent5d percent2c+2001&issue=-<=1 (accessed December 13, 2013).

7. See "Working Population by Educational Attainment (Highest Level Completed) and Occupation [Sub Major Group], 2006," *2006 By-Population Census,* http://www.bycensus2006.gov.hk/FileManager/EN/Content_981/c120e.xls (accessed December 13, 2013).

8. See "Working Population by Occupation, Sex and Educational Attainment (Highest Level Attended)," *2011 Population Census,* http://idds.census2011.gov.hk/Main_tables/Batch_04/C124.XLSX (accessed December 13, 2013).

9. See Richard Florida, *The Rise of the Creative Class: Revisited* (New York: Basic Books, 2012).

10. Ibid.

11. "Report of the Phase Two Review of the Post Secondary Education Sector," Education Bureau, 2008, http://www.edb.gov.hk/attachment/en/about-edb/publications-stat/major-reports/phase2reviewreport(eng).pdf (accessed December 13, 2013).

12. Florida et al., *Creativity and Prosperity*.
13. J. Needham, *Science and Civilization in China* (Cambridge: Cambridge University Press, 1956).
14. M. Chugani, "Critical thinking and rote learning," *Headline: Hong Kong*, November 15, 2011, 26.
15. S. F. Yip, Creativity and functionalism make a great city. *South China Morning Post: Hong Kong*, April 19, 103, http://www.scmp.com/news/hong-kong/article/1217724/creativity-and-functionalism-make-great-city (accessed December 13, 2013).
16. See E. W. Cagape, HK kids' lack of creativity blamed on parents, *Asian Correspondent,* May 2, 2012, http://asiancorrespondent.com/77226/hk-kids-lack-of-creativity-blamed-on-parents/ (accessed December 13, 2013).

Section 3: Comparative and Cross-Border Perspectives

Chapter 14

Transparent Cosmopolitanism: New Versions of the *Flâneur* in Taiwanese and Hong Kong Cinema

Joey Moon

Modernization initially started during the late nineteenth century and evolved during the twentieth century. In terms of theoretical hypothesis, modern societies that are considered civilized evolved in several ways, such as through dynamic philosophical and literary movements and the development of cognitive psychology. The rapid development of globalization in the twenty-first century represents a new world order because increasingly unified and efficient international cooperation is creating unprecedented wealth, goods, and services for some. Globalization includes economic, technological, sociocultural, political, and biological elements driven by transnational transmissions of ideas and languages (Friedman, 1994). However, antiglobalization perspectives have also been proposed because of conflicts that globalization creates between international and local cultures, particularly in developing countries. Therefore, despite my being born and raised in developing countries, I feel compelled to search for theories that more adequately explain the evolution of globalization. A review of the relevant literature shows that the theory of cosmopolitanism can fill the gap in the literature on globalization as it relates to social, political, and cultural theory. According to Delanty (2009), globalization does not have a philosophical or methodological framework beyond the basic assumptions of modern social science. He further suggested that globalization could be conceptually and philosophically assessed through the lens of cosmopolitan theory because it provides a way to account for cultural diversity within the framework of globalization and connects detailed philosophical approaches with empirical sociological demands. Beck (2006) argued that in a cosmopolitan world, distinctions, contrasts, and boundaries must be set and defined based on a consciousness of shared similarities between in and out groups. In addition, because of individual difference, boundaries among groups are becoming more fluid. The notion of fluid boundaries in cosmopolitanism reflects the everyday realities at both national and international levels. Therefore, countries have become spaces of encounters, intermixing, uncertain coexistence, and the overlapping of life-worlds and global dangers. These elements require a reevaluation of individuals' connection with the local and global. Based on these elements, Beck (2006) further explains the significance of the cosmopolitan city by analyzing the work of three cosmopolitan writers with roots in Munich who became expatriate writers. These authors created their work at a time when Munich was trying escape its painful history and reinvent its identity in line with the discontinuity of its history. Therefore, cosmopolitan Munich opened up to the world and accepted the dangers and crises produced by global civilization. During this process, former differentiations between internal and external, national and international, us and them have lost their official identities. A new cosmopolitan reality is essential everywhere for people to survive. Nevertheless, cosmopolitan cities have opened up to the world and accept the related dangers and crises. Human behavior in these types of urban environments may also develop distinct traits. The character of the *flâneur* is often used as an example of the new urban sensibility drawing from the rich diversity available in cities. The term *flâneur* dates to the nineteenth century and was first used by Charles Baudelaire (1964) to describe modern cosmopolitan individuals who were far from home yet felt at home all over. In addition, *flâneurs* were spectators who used the crowd as a lens to view the world but might select to remain hidden from the world. Furthermore, the *flâneur* could mirror the crowd itself and reflect each one of its movements, reconstructed by multiple outstanding segments from all aspects of life. Numerous scholars have used Baudelaire's *flâneur* as a crucial representative of the modern urban spectator. Among these scholars, Walter Benjamin was perhaps the most distinguished literary critic. He claimed (2006) that *flâneurs* might appear to be walking around aimlessly but in reality were actively in search of city spaces that mirrored their inner perspectives. By wandering around in a city, *flâneurs* adopt several identities, such as pedestrian, dandy, idler, gangster, gambler, and prostitute, and all these roles are performed on the city as a stage to create the illusion of an unusual, costumed self. Thus the *flâneur* has a genuinely fluid character that may lack national, ethnic, and local identities. The fluid cosmopolitan character of the *flâneur* contradicts the traditional boundaries that constitute us and them. Therefore, the primary goal of this paper is to explore to what extent the fluid character of the *flâneur* is helpful in developing a cosmopolitan understanding that can bridge the gap between local people and outsiders. These concepts will be explored in the following literature review section. Ulrich Beck (2006) identifies five characteristics of a cosmopolitan person. The first is the ability to overcome boundaries between internal and external, and national and international. The second is the ability to understand cosmopolitan differences and contradictions and to be curious about cultural dimensions and identities. The third is the cosmopolitan perspective that understands that situations have the potential to be both opportunity and risk. The fourth is recognition of the impossibility of living in a world with boundaries. The fifth is recognition of the interconnectedness of the local, national, ethnic, religious, cultural, and traditional. To summarize, a cosmopolitan person is aware of similarities shared with others and aware that they are not fixed and defined by differences, contrasts, and boundaries. Therefore boundaries between us and them are not blurred by individual particularities but instead become wide open. In addition, Beck pointed out that intellectual pioneers have been constructing this type of internalized cosmopolitanism ever since the modern era. The studies of modernist scholars therefore provide the most appropriate direction for this study.Because of the effect of cosmopolitanism on the ontology of particular individuals, the normative significance of cosmopolitanism, which consists of an alternative type of reality that can be explained by modernist theories, is explored in this paper. The goal is to reveal the existential particularities of individuals. Benjamin (Gilloch, 1996) explained that among cosmopolitan individuals, *flâneurs* in particular searched out city spaces to reflect their internal perspectives. In addition, he suggested that only photography and motion pictures were able to visually portray the significance of the city because of their ability to capture the fluid movement of the urban environment, including the activities of *flâneurs*. Nevertheless, it is clear that the twenty-first century *flâneur* does not completely comply with the descriptions of Baudelaire, Benjamin, and other scholars. For example, the character of the *flâneur* was used to explain modern urban experiences, such as the spectator's gaze, class tensions in cities, modernist alienation as defined by expatriate writers, and the development of mass culture with the crowd at its base. Therefore, reconceptualizing the contemporary *flâneur*, who is more fluid, could underpin an empirical framework for reality that could examine the existential particularities of individuals who represent the fluid boundaries between in and out groups. The following definitions that constitute the base of the analysis framework for the selected films are reviewed: cosmopolitanism; *flâneur*. The aim is to incorporate these definitions into the film analysis to ultimately explain the contexts portrayed in the films by applying these theories. This section is the core of the paper.

RESEARCH PURPOSE, QUESTIONS, AND METHODOLOGY

Reconceptualizing the contemporary *flâneur* is the main focus of this paper. Historically, Benjamin argued that the *flâneur* was a producer of texts and images. The *flâneur's* gaze was directed toward specific persons and objects. Due to the *flâneur* producing texts and images, Benjamin identified cinema as a representation of the *flâneur's* gaze (Gilloch, 1996). The following research questions are based on this perspective of cinematic representation.

1. Are audiences able to observe twenty-first century social developments through the new *flâneur* represented in Taiwanese and Hong Kong films?

2. In terms of film aesthetics, how can the new group of *flâneurs* be identified, and how do these fluid characters differ from previous versions?

3. Is the bridge between locals and others represented by the *flâneur* able to constitute cities where the possibilities of rejection, questioning, confrontation, and acceptance of the fluid characteristics of the *flâneur* can be experienced?

4. Are *flâneurs* able to construct communities with fluid boundaries that comply with cosmopolitan principles? In this study, field theory is used to analyze two Taiwanese and Hong Kong films. Field theory was selected because it is suitable for research fields that are independent, interconnected, delimited, and exposed to new connections. Each field is characterized by a central problem and by a series of branches that connect it to other fields. Each field has its own identity and is often rich in experiences,

yet has long stable bases. For instance, economic, social, political, and psychological aspects inform theories on cosmopolitanism and the *flâneur*. Each theory is interconnected and reflective of the humans that live in the field. Although Benjamin and other scholars viewed cinema as representations of the human condition, field theory allows for deeper analysis that also evaluates problems and phenomena caused by cinema.

FILM ANALYSIS AND OUTCOMES

The analysis emphasizes large segments of two films, *Sparrow* and *Parking* (both produced in 2008). Famous Hong Kong director Johnny To directed *Sparrow*. The film explores various angles on the lives of a group of pickpockets in Hong Kong and in the process reflects on social changes in Hong Kong using cinematic images. The story begins with a group of young pickpockets who fulfill *flâneur* roles. The group preys on victims by wandering around the city and in doing so connects with a beautiful girl who is controlled by the leader of another group of pickpockets. At a later stage in the film, it is revealed that she is a foreigner from mainland China—because she has a People's Republic of China passport. The young pickpockets go through a process of rejecting, questioning, confronting, and eventually accepting the beautiful Chinese girl because her life is in danger. This further emphasizes the existence and sharing of cosmopolitan sympathy by saving her from control of a local group. The locally produced Taiwanese film *Parking* was directed by Mong-Hong Chung, who has been hailed as the successor of Edward Yang in Taiwanese Cinema. Chung portrays the lives of *flâneurs* and cosmopolitan individuals in Taipei. This film delivers the important message that Taiwan is becoming a diverse country with fluid boundaries. *Parking* follows a more complex approach than *Sparrow* in its portrayal of *flâneurs*. The protagonist is forced to become a *flâneur* for a day in Taipei because he has to buy cakes for his wife. Unexpectedly, his car is blocked inside a parking lot by double-parked vehicles. In trying to find the owners of the double-parked vehicles, he meets two different local gangsters. One is a pimp for a woman from mainland China called Lee Wei. The other is a loan shark tracking down a tailor from Hong Kong to collect debts. Eventually a former gangster who now is a barber helps the tailor. Later the tailor helps him help the Chinese girl escape from her pimp. Finally, the protagonist adopts an orphan girl from her grandparents. Similar to *Sparrow*, the psychological processes the protagonist experiences are fluid and involve rejection, questioning, confrontation, and acceptance of others, and he eventually assists them in changing their lives. Additionally, he creates his own fluid community by adopting an orphan.

VALUE OF THE RESEARCH AND CONCLUSION

This study demonstrates the fluidity of *flâneur* characters in these two films, and this becomes obvious as soon as they recognize that outsiders are similar to them. In addition, it explores the bridge *flâneurs* represent between locals and outsiders, which symbolizes the contemporary city experience. Thus the fluidity of *flâneurs* allows them to experience possibilities, such as rejection, questioning, confrontation, and acceptance. These findings reveal that the new type of *flâneur* may potentially adopt the role of an antinationalist character, which could be explored in future research.

In the following literature review section, theories essential to the understanding of cosmopolitanism and the *flâneur* are reviewed. During the twentieth century, globalization rapidly expanded worldwide. The related international integration also involved economic reforms and intensified immigration. By the turn of the twenty-first century, globalization constituted a new world order, which efficiently unified labor globally to produce unprecedented levels of wealth and goods (Friedman, 1994). However, Delanty (2009) argued that globalization did not represent a philosophical or methodological framework that broadened the basic assumptions of modern social science. In addition, he suggested that the concept and philosophical basis of globalization could be explained with the concept of cosmopolitanism, which is a theory of globalization used in social, political, and cultural studies. Cosmopolitanism is appropriate because it accounts for cultural diversity in globalization and is able to connect detailed philosophical approaches with empirical sociological demands. Therefore, the following section includes a review of some of the key definitions and theories that form the basis for the analytical framework applied to the film contexts.

DEFINITIONS OF COSMOPOLITANISM

The term *cosmopolitanism* has been used in theories related to the premodern, modern, and postmodern to examine radical changes involving permanent and temporary residents who managed to establish mutual respect (Appiah, 2007). According to Beck (2006), cosmopolitanism represents a global sense of boundlessness related to daily life, history, awareness of contradictions related to social background, and blurred distinctions that establish cultural peculiarities. Cosmopolitan theory reflects on potential changes in human lifestyles and social relations when diverse cultures mix.

Beck analyzed cosmopolitanization following a multidimensional approach. He thus explained that multiple loyalties reflect increases in various transnational lifestyles that also include a reaction against globalism while being supportive of a distinct cosmopolitan globalization. However, the term *cosmopolitanization* does not fully account for long-standing and ongoing globalizing currents. After all, people have been traveling between numerous places, driven by processes related to production and consumption, for years. Therefore, in reality, relationships have been interdependent and enhancing living conditions for a long time. The dispersal of territorial communities and the spread of global, social, and economic interdependence are accounted for by cosmopolitanism, because it incorporates diverse traditional approaches and people and their interactions both nationally and internationally. Therefore, cosmopolitan places could also experience sudden conflicts and expose people and countries to multiple global risks, which demand reconceptualizations of the relationship between local and global. To summarize, cosmopolitanism indicates an awareness of shared similarities with others, contrasts, and boundaries and that these differences are neither fixed nor clearly defined. The boundaries between in and out groups are not blurred by particularities but instead have become wide open. The experience of reconfiguring boundaries constitutes the core of global risks, which demands a reconfiguration of the relationship between local and global. In addition, Delanty (2009) argues that intellectual pioneers have been constructing internal cosmopolitanism ever since the modernist era. Therefore, modernist theories provide the most appropriate direction for this study.

DEFINITIONS OF THE *FLÂNEUR*

Because of the effect of cosmopolitanism on particular individuals, the normative significance of cosmopolitanism consists of an alternative type of reality that can be explained with modernist theories. These theories discuss a new type of more fluid *flâneur* present in urban spaces. Kristeva's (1991) philosophy, for instance, contradicts orthodox psychoanalysis by stating that the ability to break free from repression represents a daring freedom. Foreigners challenge locals' sense of identity, and because they experience this as losing control, they feel wounded. In contrast to people who choose to be strangers in a strange space, *flâneurs*, as Benjamin claimed (Gilloch, 1996), may appear to aimlessly walk around but are in fact searching for a city space that reflects them. According to Frisby (2001), *flâneurs* contribute to sociological knowledge of their generation. In addition, *flâneurs* as producers are able to reconstruct the metropolitan experience of modernity and imagine beyond the visual control of metropolitan interactions. Moreover, Frisby argues that a key figure in urban culture is the stranger, that the city is a place where several strangers meet, and that these encounters are not always friendly. Diverse communities' experiences with encountering strangers provide them the opportunity to work through hatred and achieve rational and creative change. Subsequently, the city was perceived as the locus of the *flâneur*. The modernist theorist Baudelaire (1964) argued that the modern city was the only place to escape from boredom, and *flâneurs* thus became part of the modern crowd. However, theoretically the crowd was considered separate from *flâneurs*. *Flâneurs* play several contrasting roles in the city. On the one hand, they play characters, such as pedestrians, dandies, or idlers, that could be true representatives of cities. On the other hand, they also play the roles of gangsters, gamblers, and prostitutes, who have the potential to ruin cities. The imaginative approach features the fluid character of metropolitan existence in modernity and rejects a systematic outlook (Gilloch, 1996).In short, cosmopolitanism has five characteristics, which represent a globalizing perspective on the contemporary human condition. Nevertheless, within the framework of the cosmopolitan, it is clear that twenty-first-century *flâneurs* have moved beyond the definitions of Baudelaire, Benjamin, and other scholars. For example, the character of the *flâneur* explains the modern urban experience and gaze of the spectator, class tensions in cities, modern alienation as defined by expatriate authors, and the crowd as basis for mass culture. Reconceptualizing the new *flâneur*, who is more fluid, could involve an empirical framework of reality and examine the existential particularities of individuals who act as fluid boundaries between in and out groups. The cinematic perspective of Benjamin proposes that only films, as a visual method, are able to focus on the significance of current movements and the situation of *flâneurs* in cities. The research purpose and questions are illustrated in the next section. In addition, the study methodology is explained based on the research purpose and questions.

RESEARCH PURPOSE, QUESTIONS, AND METHODOLOGY

This study aims to reveal that among cosmopolitan people, a new group of *flâneurs* can be identified. They repeatedly expose themselves to a wide range of people and foreign places and do not feel at home anywhere. Benjamin pointed out that motion pictures are able to feature the fluid movements of immigrants and *flâneurs* in cities. This assertion directs this study on the fluid movements of immigrants and *flâneurs* in cities in two motion pictures.

Research Purpose and Questions

Relevant literary theories were reviewed, and the catalog of Taiwanese cinema from 2000 onward was examined. Based on these theories, films were selected for screening and examination. The following questions were considered:

1. Are audiences able to observe twenty-first-century social changes through the new *flâneur* represented in Taiwanese and Hong Kong films?

2. In terms of film aesthetics, how can the new group of *flâneurs* be identified, and how do these fluid characters differ from previous versions?

3. Is the bridge between locals and others represented by the *flâneur* able to constitute cities where possibilities of rejection, questioning, confrontation, and acceptance of the fluid features of the *flâneur* can be experienced?

4. Are *flâneurs* able to construct communities with fluid boundaries that comply with cosmopolitan principles? Field theory was used to analyze the selected films. It was selected because it is suitable for research fields that are independent, interconnected, delimited, and exposed to new connections.

Methodology

Identifying the new group of *flâneurs* was the primary aim of this study. Benjamin argued that *flâneurs* were historically producers of texts and images in cities. The *flâneur's* gaze was directed toward particular persons and objects. Due to the *flâneur* producing texts and images, Benjamin identified cinema as representation of the *flâneur's* gaze (Gilloch, 1996). Therefore, film contexts are considered as reflecting social reality in this study. Based on a review of the literature on cosmopolitanism and *flâneurs*, research questions were proposed from the cinematic representation perspective. Linked to the theory of cosmopolitanism, the particularities of individuals are the most essential principle. Therefore, this study aims to analyze new types of *flâneurs* in contemporary Hong Kong and Taiwanese films. The methodology is based on Cassetti's cinematic theories developed from 1945 to 1995 (Wu, 2008). Three paradigms have been applied to film research strategies, including ontology, methodology, and field theory. This study uses the field theory method because it is able to account for interconnected research fields. Each field is characterized by a central problem and a series of branches that connect it to other fields. Each field has its own identity; it is often rich in experiences yet has long stable bases. For instance, theories of cosmopolitanism and *flâneurs* are informed by economic, social, political, and psychological factors. Each theory is interconnected and reflects the human condition. Although Benjamin and other scholars considered cinema as a representation of human lives, field theory also allows researchers to focus on problems and phenomena caused by cinema. Consequently, elements of field theory were applied as tools to analyze the film contexts described in the next section.

FILM ANALYSIS AND OUTCOMES

This study demonstrates the fluidity of *flâneurs* in the selected films, and this becomes obvious as soon as they recognize that outsiders are similar to them. Two films portraying this reality are *Sparrow* by Johnnie To and *Parking* by Mong-Hong Chung (both produced in 2008). The *flâneur* characters in these films could be divided into two social groups—namely, locals and outsiders, who are mainly from China. They act as bridges to redefine the fluid communities of contemporary Taiwan and Hong Kong, and they also articulate connections and contradictions between Taiwan, Hong Kong, and China. In addition, this study explores the bridge that *flâneurs* represent between locals and outsiders, which symbolizes the contemporary city experience. Thus the fluidity of *flâneurs* allows them to experience possibilities, such as rejection, questioning, confrontation, and acceptance.

Sparrow (2008)

Sparrow was produced by famous Hong Kong director Johnny To, who portrayed different angles of a group of pickpockets living in Hong Kong. In the process, he portrayed social change in Hong Kong using cinematic images. The story begins with a group of young pickpockets who fulfilled *flâneur* roles. The group preys on victims by wandering around the city and thus connects with a beautiful girl, who is controlled by the leader of another group of pickpockets. At a later stage in the film, it is revealed that she is a foreigner from mainland China, because she has a People's Republic of China passport. The young pickpockets go through a process of rejecting, questioning, confronting, and eventually accepting the beautiful Chinese girl, because her life is in danger. This further emphasizes the existence and sharing of cosmopolitan sympathy by saving her from the control of a local group.

The film is constructed as follows:

1. *Flâneurs* = a group of young local pickpockets with fluid characteristics but empathy for victims.

2. The *flâneur* = the foreign other = a beautiful girl wandering the streets looking for help.

3. *Flâneurs* = local older pickpockets driving around the city to keep an eye on the girl.

4. The trigger of cosmopolitan sympathy = the People's Republic of China passport of the beautiful girl found on the leader of the older pickpockets, Mr. Fu. Before the cosmopolitan sympathy was triggered, the local young pickpockets rejected, questioned, and confronted the beautiful girl, Chung Chun Lei, for injuring them. In addition, when they confronted the older pickpockets, their common cosmopolitan sympathy was triggered to save Chung from Mr. Fu. This film, as well as numerous other Hong Kong productions after the 1997 handover, explored the ambiguous relationship between China and Hong Kong. The beautiful girl from mainland China, Chung Chun Lei, explains, "Mr. Fu treats me well, very well, and I shall accompany him till the day he dies." Later, when she asks the young pickpockets, Kei, Bo, Mac, and Sak, for assistance to escape from Mr. Fu, she says, "No, I shall find someone I love and that I truly love to be with." It also symbolizes the love/hate relationship between China and Hong Kong. In addition, in order to portray a common shared sympathy among all Johnny To's film characters through a sense of humor and justice linked with traditional Chinese masculinity. These sentiments inform the core aesthetic of all his films. Particularly, he portrayed

social changes in Hong Kong with fluid photography, signaling the introduction of a novel type of *flâneur* in this film. Therefore, To revealed that the new *flâneur* was not aimlessly walking around the city but was able to create a fluid cosmopolitan world and wandered around urban areas for specific reasons. However, based on Cheung's (2011) perspectives, several Hong Kong directors combine a cosmopolitan understanding of film form with a realist commitment to the particularities of local identities. To is among the few Hong Kong directors who are able to portray the self–other relationship in a cosmopolitan context while exploring the ethics of generosity.

Parking (2008)

The locally produced Taiwanese film *Parking* was directed by Mong-Hong Chung, who has been hailed as the successor of Edward Yang in Taiwanese cinema. In this film, Chung investigates the lives of *flâneurs* and cosmopolitan individuals in Taipei and delivers the important message that Taiwan is becoming a diverse and fluid community. *Parking* follows a more complex approach than *Sparrow* in portraying a group of *flâneurs*. The protagonist is forced to become a *flâneur* for a day in Taipei because he has to buy cakes for his wife. Unexpectedly, his car is blocked inside a parking lot by double-parked vehicles. In the process of finding the owners of the double-parked vehicles, he meets two different local gangsters. One is a pimp for a woman from mainland China called Lee Wei. The other is a loan shark tracking down a tailor from Hong Kong to collect debts. Eventually, a former gangster who now is a barber helps the tailor. Later the tailor helps him help the Chinese girl escape her pimp. Finally, the protagonist adopts an orphan girl from her grandparents. Similar to *Sparrow*, the psychological process the protagonist experiences is fluid and involves rejection, questioning, confrontation, and acceptance of others, and he eventually assists them in changing their lives. Additionally, he creates his own fluid community by adopting an orphan. The construction of this film is:

1. The *flâneur* = local protagonist trying to buy cakes for his wife.

2. The *flâneur* = foreign other = Lee Wei, a sex worker from mainland China. She tries to escape her Taiwanese pimp, Ah Bao.

3. The *flâneur* = foreign other = a Hong Kong tailor who tries to hide from loan sharks who are collecting debts.

4. Trigger of cosmopolitan sympathy = Ah Bao and the loan sharks' violent tendencies—they beat people up.

5. In this film, Chung presents several psychological layers of the protagonist, such as rejection, questioning, confrontation, and acceptance. The protagonist appears to be a member of the Taipei bourgeoisie; to transform from ignorant *flâneur* to compassionate cosmopolitan, he encounters foreigners and other classes of local people. His psychological processing also symbolizes the social change of encountering outsiders from China and Hong Kong that Taiwan has been experiencing since 1997. Chung explains through this film that the relationships among China, Hong Kong, and Taiwan have become more complicated. As explained by the Hong Kong tailor, "I noticed that my dad's tailor shop was not doing well in Hong Kong, and he therefore moved to Taiwan. I followed him after my mum died. However, my dad's business was still not doing well in Taiwan. After he died, I went to China, and although I was cheated once in China, I at least knew that it would be a new beginning there." The Chinese sex worker, Lee Wei, explores a similar situation when she asks her pimp, "Where is my money? Why did you say that I only needed two or three months to pay off all my debts? Now, I still need to work another two or three months?" The pimp will not answer her questions and she is forced to escape to find her own truth. Because of the ambiguous relationships among Taiwan, Hong Kong, and China, Chung makes his audience aware of differences between in and out groups through the protagonist who encounters both local people and foreigners. The help he receives from these people triggers a common sympathy in him and results in the construction of a fluid cosmopolitan community. Chung also portrays social changes in Taipei with cinematography and setting. The orphan girl and her grandparents are victims of these social changes. The sex worker from mainland China, the Hong Kong tailor, and their families are also affected by social changes in China and Hong Kong within the borders

of mainland China. Two of these characters are forced to become *flâneurs* in Taipei because of their work situation and debts. Eventually, they are also forced to be *flâneurs* elsewhere in Taiwan or in different Chinese cities. Chung clearly aims to highlight the importance of raising awareness of individual and cultural diversity in Taiwanese society as it has become more open to an influx of people from Hong Kong and mainland China. It is therefore essential that Taiwanese society adopt a fluid cosmopolitan outlook.

To summarize, Taiwan and Hong Kong have been experiencing radical social changes during the twenty-first century, particularly with regard to their relationships with mainland China. These changes are explored by identifying a new group of *flâneurs*. Because of the contrasts among the *flâneurs,* locals, and others, cities where these characters might experience rejection, questioning, confrontation, and acceptance become possible through the fluid character of the *flâneur*. Therefore, the new group of *flâneur* is able to construct communities with fluid boundaries in line with the principles of cosmopolitanism. Consequently, the empirical framework provided by field theory is used to explore the new type of *flâneur* who manages to bridge the dichotomy between locals and outsiders. The film analysis reveals that they are instrumental in creating novel cosmopolitan narratives in response to radical social changes in Taiwan, Hong Kong, and mainland China. The value of the research is discussed and a conclusion is provided in the next section.

VALUE OF RESEARCH AND CONCLUSION

Value of the Research

This study demonstrates the fluidity of *flâneur* characters in these films, and this becomes obvious as soon as they recognize that others are similar to them. In addition, it explores the bridge *flâneurs* represent between locals and outsiders, which constructs the contemporary city experience. Thus the fluidity of *flâneurs* allows them to experience possibilities, such as rejection, questioning, confrontation, and acceptance. The empirical framework provided by field theory is used to explore the new type of *flâneur* who manages to bridge the dichotomy between locals and outsiders. These characters are instrumental in creating novel cosmopolitan narratives in response to radical social changes in Taiwan, Hong Kong, and mainland China. This study identifies a new type of *flâneur* with a more fluid character, who is able to construct fluid cosmopolitan communities in urban spaces. In addition, the findings reveal that the new type of *flâneur* has potentially transformed his or her role into one with an antinationalist character, which could be explored in future research.

Concluding Observations

Globalization has introduced a situation where economies and consumption transgress international borders. However, the crisis Western capitalism (regarded as the driving force behind globalization) has been experiencing during the twenty-first century has meant that the term *globalization* was not applicable when Western countries faced economic recession. Rather than the globalization of transnational enterprises, a transpersonal globalization known as cosmopolitanism has been foregrounded. Cosmopolitanism is based on overcoming various boundaries. Cosmopolitan individuals understand difference, contradictions, and the diverse dimensions of culture and identity. They recognize that any situation could be either opportunity or risk and that living in a world with boundaries is impossible. Ultimately, cosmopolitanism is based on the interconnectedness of the local, national, ethnic, religious, cultural, and traditional. In the cosmopolitan era, new groups of *flâneurs* are forced to wander the globe and are further selected as an empirical framework to examine the contemporary reality and existential particularities of individuals who manage to construct fluid boundaries between in and out groups. Benjamin's cinematic perspective still applies because he suggested that the visual approach of motion pictures was able to focus on the significance of continuous movements and of the situation of immigrants and *flâneurs* in cities. The analysis of the films provides clear answers to the research questions. It reveals that the leading roles of fluid characters, such as the *flâneur,* represent a dichotomy between locals and outsiders that facilitates the construction of novel narratives on cosmopolitanism in response to radical social changes in Taiwan, Hong Kong, and China. In the two films analyzed in this study, audiences can easily identify with the new type of *flâneurs* by being exposed to their experiences with rejection, questioning, confrontation, and acceptance. By the end of both films, the *flâneurs* are able to construct their own fluid communities from the perspective of their own fluid internal boundaries. However, this research has limitations. It may be difficult to find more examples of Hong Kong and Chinese films with cosmopolitan themes, particularly films with *flâneur* leading characters.

In addition, the small sample of only two films may compromise the generalizability of the findings. Furthermore, the scope of the research would have been too big if additional comparisons between films and social changes had to be made. Future studies could explore more literature from the twentieth century to contemporary times. In addition, they could explore new directions that this type of *flâneur* is likely to take, such as becoming an antinationalist character. Following this approach, more cinematic models from other Asian cinemas could be incorporated into the empirical framework. Moreover, the methodology could be tested in a variety of ways to identify additional angles and contexts in diverse media.

BIBLIOGRAPHY

Ahern, M. 2003. "Gilles Deleuze: Nomad Thought." *Theories of Media* (winter), http://csmt.uchicago.edu/annotations/deleuzenomad.htm.

Appiah, K. A. 2007. *Cosmopolitanism: Ethics in a World of Strangers*. New York: W. W. Norton.

Baudelaire, C. 1964. *The Painter of Modern Life and Other Essays*. London: Phaidon.

Beck, U. 2006. *Cosmopolitan Vision*. Cambridge: Polity.

Benjamin, W. 2006. *The Writer of Modern Life: Essays on Charles Baudelaire*. Cambridge, MA: Belknap Press of Harvard University Press.

Butler, J. 2004. *Precarious Life: The Powers of Mourning and Violence*. London: Verso.

Cheung, E. M. K., G. Marchetti, and S. Tan, eds. 2011. *Hong Kong Screenscapes: From the New Wave to the Digital Frontier*. Hong Kong: Hong Kong University Press.

Delanty, G. 2009. *The Cosmopolitan Imagination: The Renewal of Critical Social Theory*. Cambridge: Cambridge University Press.

Deuchars R. 2011. "Nomad Thought, Global Capital and Political Resistance: Deleuze and Guattari's War Machine." International Studies Association Annual Convention, March 16–19, http://www.allacademic.com//meta/p_mla_apa_research_citation/5/0/2/6/0/09/ pages5026/p502609-1.php (accessed April 24, 2011).

Eisenstadt, S. N. 2002. *Multiple Modernities*. New Brunswick, NJ: Transaction Publishers.

Friedman, J. 1994. *Cultural Identity and Global Process*. London: Sage.

Frisby, D. 2001. *Cityscapes of Modernity*. Cambridge: Polity.

Gilloch, G. 1996. *Myth and Metropolis: Walter Benjamin and the City*. Cambridge: Polity.

Gilroy, P. 2004. *After Empire: Melancholia or Convivial Culture?* Abingdon, UK: Routledge.

Held, D. 2010. *Cosmopolitanism: Ideals and Realities*. Cambridge: Polity.

Kristeva, J. 1991. *Strangers to Ourselves*. New York: Columbia University Press.

Lash, S., and J. Friedman, eds. 1992. *Modernity and Identity*. Oxford: Blackwell.

Vertovec, S., and R. Cohen, eds. 2008. *Conceiving Cosmopolitanism: Theory, Context, and Practice*. Oxford: Oxford University Press.

Westwood, S., and J. Williams, eds. 1997. *Imaging Cities: Scripts, Signs, Memories*. London: Routledge.

Wu, I., trans. 1996. *L'Analyse des Films*. Taipei: Bookman Publishing.

Wu, I. 2008. *In the Era of Cinematic Thinking*. Taipei: Bookman Publishing.

Chapter 15

Educating Entrepreneurial Portfolio Musicians for Twenty-First-Century Careers

Patrick M. Jones

Today's musicians operate in an environment unlike any before. They have instant access to digitized musical and cultural products and artifacts from around the world and access to other musicians and worldwide audiences. The ability to share their work with a global audience via the Internet; communicate with others through cheap, simple, and near-instant means; and travel easily and affordably has made sharing, learning, borrowing, mimicking, recontextualizing, and repurposing musical and cultural works, performances, genres, forms, and ideas easier than ever before. One result of this digital revolution is that mediating organizations such as corporations, governments, and record labels no longer exercise the same degree of control they did up until the late twentieth century. It is a new era with new roles, rules, and risks. Musicians must exercise a degree of self-management for which they have not traditionally been prepared. In short, they must operate as entrepreneurs. They must be outstanding, flexible, multifaceted, and entrepreneurial. They must have the skills and knowledge to maintain a portfolio career, which involves earning revenue through a variety of musicianly roles and activities, including performing, composing, arranging, recording, teaching, starting their own ensembles, creating new products, finding new audiences, and managing their own business affairs.

This is not the first time in history that musicians have had to be entrepreneurs with portfolio careers. A survey of Western classical-music history reveals that musicians in earlier periods also worked this way. Pamela Starr identified three types of entrepreneurial career paths for musicians in the fifteenth century based on the careers of four particular musicians. Those three paths are not dissimilar to those followed today. They include moving through a series of positions of increasing prestige and income throughout one's career, establishing a stable position and supplementing one's income with additional outside work, and working primarily as an administrator or teacher of music while pursuing one's performing or composing interests.[1] The Baroque musician Johann Sebastian Bach is a perfect example of a portfolio musician. He held steady positions as a court and church musician, composed and arranged major works and music for weekly church services, taught music lessons, and was a conductor and an outstanding performer who knew how to market himself.[2]

We have the ability to help music majors prepare for such careers. We simply need to provide the appropriate offerings. In addition to course work and other experiences, we should prepare students for the global nature of the music industry. Institutions should collaborate to prepare musicians for a global music workplace with worldwide opportunities and challenges, varying regulations and working conditions, and audiences with a multitude of cultural heritages and aesthetic expectations.

In this paper I review research on musician careers, discuss initiatives on educating musicians to be entrepreneurs with portfolio careers, provide examples of our efforts at the Syracuse University Setnor School of Music, and discuss ways that international collaborations and partnerships prepare musicians for global audiences and conditions.

MUSICIAN CAREERS IN THE TWENTY-FIRST CENTURY

Two terms currently popular regarding the work of professional musicians are *entrepreneur* and *portfolio career*. Entrepreneurs are typically understood to be people who start businesses and create products and services to meet needs. A portfolio career involves making a living by combining a variety of income streams from various jobs, such as performer, composer, teacher, and manager. While these roles are not interchangeable, they are complementary for working musicians, who may need to be entrepreneurs in addition to or as a direct result of having portfolio careers. A review of current literature reveals that most working musicians in the United States make a living through portfolio careers. Graduates of leading arts institutions and working professional musicians were surveyed for two recent research projects on arts careers.

The results of both surveys indicate that portfolio careers predominate among all artists in general and among working musicians in particular.

A major initiative to get a sense of the work of arts graduates is the Strategic National Arts Alumni Project, based at the Indiana University Center for Postsecondary Research. Its 2010 survey of arts alumni from U.S. institutions received responses from 13,581 alumni of 154 institutions. The survey revealed that "63% of arts graduates were self-employed at some point in their career. They are entrepreneurial, with 14% having started their own company or organization. And, consistent with a popular view of the professional artist, almost three-fifths (57%) of current professional artists hold two or more jobs concurrently."[3]

The Future of Music Coalition surveyed U.S.-based musicians and composers from September 6 to October 28, 2011. While 4,652 respondents completed the entire survey, 5,371 completed the central questions. The survey revealed that 82 percent of respondents earned their music-related income from activities in two or more roles in the twelve-month period preceding the survey, as illustrated in Figure 15.1.[4]

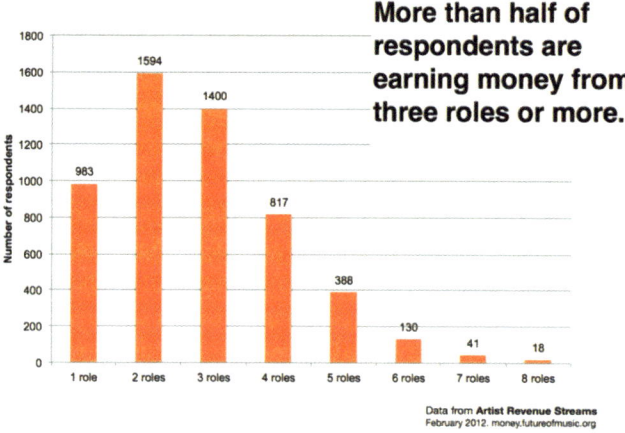

Figure 15.1 Number of roles for music-related income

The research on revenue streams of the 2,728 respondents self-identifying as conservatory or music school graduates revealed that they derived their incomes from eight different sources, including teaching, performing, composing, and other streams, such as recordings and merchandise branding. Their income is aggregated and displayed in a pie chart in Figure 15.2.

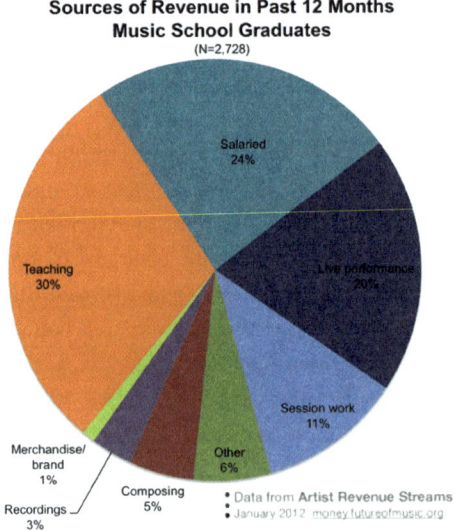

Figure 15.2 Aggregated revenue streams of conservatory of music school graduates

Thus 57 percent of all working artists and 82 percent of musicians responding to the two surveys made their living through a portfolio of roles. The music school graduates also reported that they derived their income from eight different sources of musical work.

This situation is not unique to the United States. Research from Europe indicates that musicians there also cobble together a living by performing a variety of musical activities and roles.[5] For example, in 2008 over half (69 percent) of Dutch professional musicians reported having some type of portfolio career that combined a variety of music jobs.[6] The Royal College of Music interviewed more than thirteen hundred alumni to determine how they made a living. More than 93 percent of them were still professionally active as musicians, with most "often building their professional life from a variety of activities in several different areas."[7]

Clearly, making a living through a portfolio of musical roles and activities is not isolated to the United States. I venture to say that the majority of musicians around the globe probably make their living this way, and probably always have.

EDUCATION FOR ENTREPRENEURSHIP AND PORTFOLIO CAREERS

Musicians being entrepreneurial and having portfolio careers is not new. What has changed in recent years is the recognition of this fact and recognition that perhaps conservatories and music schools should do more to help students prepare for such careers instead of focusing strictly on preparing them for orchestral auditions or careers as soloists. Clear signs that tertiary institutions have identified entrepreneurship education as important are the rise of a cadre of professional career counselors and music entrepreneurship educators and the development of institutes, courses, programs, and even minors in music entrepreneurship in at least thirty-one American conservatories and universities, in addition to courses, programs, and institutes that include music along with other arts or disciplines.[8] The professionalization of this area is ongoing and has resulted in the creation of a variety of professional associations and an increase in scholarly activity. An association focused on arts entrepreneurship is the Entrepreneurship in the Arts Interest Group within the United States Association for Small Business and Entrepreneurship.[9] Groups specifically focused on those teaching music entrepreneurship in tertiary institutions include the College Music Society's Committee on Music Entrepreneurship Education[10] and the Arts Entrepreneurship Educators Network.[11] There is also a new peer-reviewed journal for arts entrepreneurship educators called *Artivate: A Journal of Entrepreneurship in the Arts*.[12]

Since offering entrepreneurship education to musicians is relatively new, there are a variety of approaches to doing so. The College Music Society's Committee on Music Entrepreneurship Education held a summit in January 2010 to address this topic. The papers from that conference were published as a book called *Disciplining the Arts: Teaching Entrepreneurship in Context*.[13] Ten of the book's chapters are devoted to how to teach entrepreneurship. The same group developed a handbook to help institutions start programs in entrepreneurship education for musicians.[14] Approaches include one-off workshops and lectures, workshop and lecture series, single courses, course series, organized minors, entire degree programs, entrepreneurial content embedded within existing courses, internships, and experiential learning opportunities.[15] Needs of current students can be addressed through incorporating a variety of these approaches with little or no additional expense. Many university schools of music already offer a plethora of relevant opportunities for students. The answer may simply involve making the need for such preparation explicit to students, foregrounding opportunities through creating pathways and advising sheets, and removing any institutional cultural or structural barriers to students availing themselves of the opportunities.

Even without founding institutes and formal structures, there are things institutions can do to help students prepare to be entrepreneurs. The Network of Music Career Development Officers (NETMCDO), a group of 130 career counselors in U.S. conservatories and music schools, developed recommendations for preparing students for entrepreneurial careers.[16] The recommendations are listed in Table 15.1.[17]

Table 15.1 Undergraduate learning/ teaching recommendations for career development in music

1. Life skills: Foundations of personal independence Students have begun an ongoing, introspective process of defining the strengths, values, and goals that will allow them to create, recognize, and act on fulfilling personal and professional opportunities. This area includes self-awareness, citizenship, personal development, time management, and study skills.
2. Entrepreneurial skills: The tools for implementing careers Students focus on the process of identifying and realizing opportunities while developing essential business management skills: budgeting, promotion, planning, and audience building.
3. Experiential skills: Fundamentals for navigating the creative economy Students engage with their communities and the professional world, connecting through active learning and building confidence, professionalism, and cultural and global awareness.

There are also two books available in the popular press in the United States to help musicians develop their careers. Angela Myles Beeching, director of the Center for Music Entrepreneurship at Manhattan School of Music, wrote *Beyond Talent: Creating a Successful Career in Music*[18] and maintains a complementary website (http://angelabeeching.com). David Cutler, coordinator of music entrepreneurship studies at Duquesne University, published *The Savvy Musician*[19] and also maintains an accompanying website (SavvyMusician.com). The topics of these two books can be combined into the following list:

- Audience development and engagement

- Branding (creating an image)

- Entrepreneurial mind-set
- Self-management as opposed to having an artist manager
- Financial management
- Fund-raising
- Marketing (including print and Internet materials, including social media)
- Networking
- Performing, composing, and arranging at the highest level, including freelancing
- Projects (forming ensembles, presenting concerts, commissioning works, creating recordings)
- Recording
- Teaching
- Additional employment in arts management or the music industry
- Additional self-employment, such as in music technology as an instrument retailer or repair person

SYRACUSE UNIVERSITY INITIATIVES

The Syracuse University Setnor School of Music can serve as a case study of an institution beginning to embrace responsibility for preparing students to make a livelihood in music and not simply to take auditions. The Setnor School of Music is a comprehensive university school of music with four departments: Applied Music and Performance; Music Composition, Theory, and History; Music Education; and Music and Entertainment Industries. We already offer several relevant courses and possess much of the expertise needed to help students prepare for their careers. What we have lacked is an emphasis on career development, guidance on how to prepare to be entrepreneurs with portfolio careers, some specific courses and curricular pathways, and institutional structures to focus energies and make the development of career skills explicit.

When addressing this area, it is important to recognize the difference between educating someone for entrepreneurship and educating him or her for a portfolio career. Entrepreneurship education is about developing soft skills and an entrepreneurial ethos, whereas preparing for a portfolio career requires development of discrete knowledge and skills in a variety of musical activities. Therefore, we address our students' development in both areas through both curricular and cocurricular means.

Curriculum is the core of schooling. It is through course work and curricular expectations that much of the knowledge and many of the skills students need to be successful can be developed. The following list is in addition to the traditional areas of musicianship, performance, and musical knowledge. The additional skills and knowledge students need are built on a firm foundation of musicianship, not in lieu of it. We are addressing career preparation through the curriculum in two ways. One is to add components and expectations to existing courses. The other is to add additional courses.

Adding components and expectations to existing courses can be done in ways that complement and strengthen courses. Following are only a few examples of what might be done:

- Capstone projects: Requiring students to develop business plans for existing capstone projects:

    - Based on an internship placement, students develop a business plan for starting their own nonprofit or for-profit music company, including a needs assessment, a menu of products and services, a marketing plan, and a budget.

    - Students develop a business plan for a recital, including a vision/artist statement, a marketing plan, a budget (including costs of in-kind services provided by the institution and volunteer collaborators), and assessment metrics. It could even include a touring plan, identifying additional venues and costs.

    - Students develop a business plan for starting their own music school or a traveling student-teaching service, including a needs assessment, a catalog of offerings, a marketing plan, and a budget.

- Guest lecturers: Adding guest lecturers on the topic of career development to existing lecture series

- Guest artists: Having guest artists speak about their careers in music, not just about their music

- Music history courses: Including content on how musicians once made a living as composers and/or performers

- Music theory: Having students arrange and adapt compositions for a variety of ensembles and settings

Developing new elective courses that serve the music-student population utilizes existing expertise in new ways. In our situation, each of our four departments is tasked to provide service courses for the other three departments, as follows:

- Department of Applied Music and Performance: Courses on performing in a variety of settings, such as with a studio orchestra or a pit orchestra

- Department of Music Composition, Theory, and History: Courses in arranging, orchestration, scoring, publishing, and so forth

- Department of Music Education: A course on teaching that includes pedagogy, managing a private studio, and working in teaching settings such as community music schools, community centers, prisons, retirement communities, and music stores

- Department of Music and Entertainment Industries

    - A course on the music business with content such as marketing, finance, product development, project development, business plans, and fund-raising

    - A course or courses on recording and live sound reinforcement from the performer's side of the microphone

While knowledge and skills are developed in courses, a great deal of learning happens outside the curriculum. This is particularly true for entrepreneurial skills. Therefore, we have founded or are developing the following four offerings to help students apply their knowledge and skills in order to grow them:

- Arts management minor: The arts management minor provides courses and practical experience for students to learn artist management, concert promotion, venue management and operations, finances, and philanthropy for the arts. It will include courses and students from across the College of Visual and Performing Arts and will be run by an existing full-time staff member responsible for operations of the School of Music facility.

- Career Development Center: The Career Development Center offers career advising, lectures, and workshops and provides individualized assistance with contracts, website development, résumé writing, and using social media. It is led by a part-time faculty member with an additional administrative appointment as director of the Career Development Center.

- Community Music Division: The Community Music Division is a noncredit music school open to the public. It provides opportunities to develop teaching skills in individual and group settings with students of all ages. The division worked in preliminary stages in the 2011–2012 academic year and launched fully in the 2012–2013 academic year. It is led by an existing full-time staff member whose job was redefined to include this as 50 percent of her responsibilities.

- Gilbert Week: Gilbert Week is a one-week immersion trip to New York City for music majors to meet with arts professionals, visit performing arts institutions behind the scenes, and learn about a wide variety of career opportunities in music. Students meet with people working in areas such as advancement, education, management, marketing, operations, and patron relations with leading concert venues, major music organizations, music publishers, and instrument manufacturers. The program is run by an existing full-time staff member with years of experience in arts management and operations in New York City.

Finally, extracurricular opportunities for students to develop entrepreneurial skills should not be ignored. There are a variety of student-led groups, such as presenting and performing ensembles, student chapters of professional associations, student record labels, and fraternities and sororities. These organizations can be guided to ensure that students are learning and employing best practices in leadership, project and product development, marketing, project management, budgeting, and so forth. We have a variety of such organizations in the Setnor School of Music and will be working on professional development for them in the next year.

Thus we are attempting to help our students develop as entrepreneurial portfolio musicians through adding to existing courses, developing new electives, creating institutional structures to provide cocurricular opportunities, and providing professional development to extracurricular organizations. To get started, one needn't have the funding to hire a full-time career counselor, found a fully operational entrepreneurship education center, or create new minors or degree programs. Small steps based on existing programs, with existing faculty expertise, when focused and utilized appropriately, may provide the perfect starting place. We at Setnor are just beginning and will be happy to share our experiences along the way.

INTERNATIONAL COLLABORATIONS AND PARTNERSHIPS

The global marketplace for musical products and services provides opportunities for musicians to make careers in ways that were previously unimaginable. Musicians have always exported products, such as recordings and touring performances, and collaborated with international partners. Not only are these traditional approaches easier and cheaper than before, but new opportunities for deeper cultural interaction are now available. One can now come into contact with musical and cultural products and artifacts from around the world and also make direct contact with potential audience members when sitting at home on the computer. The potential impact on local cultures and the opportunities for worldwide collaboration are immense. However, so is the potential for creating misunderstandings and engaging in misappropriation and cultural imperialism. This situation requires musicians to acquire a degree of cultural understanding and sensitivity that can be developed only by spending time in the particular society with which one hopes to engage. This is where international study, collaborations, and partnerships can play an invaluable role.

Syracuse University is one of only three American universities that run semester-long study-abroad programs for music majors.[20] Our programs include

A semester in London for music-industry majors. There is no institutional partner. Students take classes at the Syracuse University London Center and do internships with UK companies.

- A semester in Strasbourg, France. Our institutional partner is the Conservatoire de Strasbourg. Students take courses at the SU Strasbourg Center and take lessons and join ensembles at the Conservatoire de Strasbourg each spring semester. A faculty member goes for the entire spring semester. We also have short-term faculty exchanges in the form of recitals and master classes.

- A weeklong trip to Brazil each even-numbered year. It is a tour with concerts and outreach to local community groups.

Only our program in Strasbourg is an international partnership, but there are at least three types of international collaborations and partnerships possible for music schools. A robust institutional partnership would consist of all three. This might work as follows:

- Full-term student exchange: Syracuse University students attend a Chinese partner institution each fall semester. Chinese students attend SU each spring semester.

- Short-term faculty exchange: An SU faculty member gives a recital or lecture at a Chinese partner institution each spring semester on odd years. A Chinese faculty member gives a recital or lecture at SU each fall semester on even years.

- Collaborative projects: Chinese and SU students collaborate on a performance or recording project. An SU ensemble tours China based out of a Chinese institution. A Chinese ensemble tours the United States based out of SU.

Designing international programs to prepare musicians for global careers requires more than a simple exchange in which students do overseas what they would be doing at home. It can be facilitated by having them study in overseas institutions with local faculty members, perform with local musicians in the host country, learn about laws and structures of the music industry in the host country, and engage with local audiences and community members in ways that foster the development of cultural understanding. For example, students should study the local language before and during their stay abroad, should live in home stays, and should join local music organizations such as community bands, choirs, and orchestras. This would immediately put them in contact with local residents interested in music and could lead to further interactions. Their course work could include local musical content and ethnographic experiences with musical cultures in the community. Study-abroad programs should include excursions to places of cultural or musical interest in the host country that help students learn more about music within the context of the host nation. Beyond such trips, partner institutions can do a great deal of collaboration with faculty and student exchanges and joint projects. This kind of extended collaboration, working with faculty members and students from the same overseas institution throughout the time a student is enrolled in the university, will create depth of exposure and contact that can help ensure that a semester of study abroad is not simply a one-off vacation.

IMPLICATIONS FOR THE GREATER CHINA REGION

There are two major implications of this paper for the Greater China Region. One is ensuring that Chinese music majors are prepared for careers in an expanded regional and global creative economy. Governments in the Greater China Region are increasing professional cultural offerings and are focusing on growing the region's creative economy. This, coupled with the burgeoning middle class in the People's Republic of China, will cause greater numbers of students to pursue music studies. This will result in a surplus of graduates versus positions throughout the region. These graduating musicians will need entrepreneurial skills to sustain themselves. Doing so will require them to develop audiences both within and beyond the Greater China Region. Therefore, they will need both entrepreneurial skills and the ability to work with Western audiences. Developing entrepreneurial knowledge, skills, habits, and dispositions will be fostered through the kinds of curricular offerings outlined in this paper.

The second, and more challenging, implication of this paper is that Chinese musicians must be prepared to engage the West in ways that are both understandable and attractive to Western audiences. Western culture is in a state of transition at the very time China and Chinese culture are becoming more influential. Western culture is different in ways much deeper than simply artistic expression. Western thinking is based on philosophical perspectives stemming from ancient Greece and formalized through the Christian Church. This philosophical perspective is based on the premise of Cartesian dualism, where mind and body are separate. Eastern thought, however, is based on a holistic understanding without such a division. The difference has been described as Western thought focusing on "being" while Eastern thought focuses on "becoming."[21] Eastern perspectives resonate with contemporary Western concerns for ecology and the need to build a peaceful world. Thus it is a good time for Chinese and Western musicians to work together. Chinese musicians have the opportunity to help advance Eastern thought through compositions and through broadcast, live, and recorded performances for Western audiences. Understanding the West will require Chinese musicians to understand Western culture, not simply know about it. Developing the kinds of institutional collaborations and partnerships outlined in this paper is critical for helping Chinese musicians develop this kind of deep cultural understanding so they can engage Western audiences and advance and grow the creative economy of the Greater China Region.

NOTES

1. Pamela F. Starr, "Musical Entrepreneurship in 15th Century Europe," *Early Music* 32, no. 1 (2004).

2. Christoph Wolff et al., "Bach," *Grove Music Online, Oxford Music Online*, http://www.oxfordmusiconline.com.libezproxy2.syr.edu/subscriber/article/grove/music/40023pg10 (access date June 2014).

3. SNAAP, "Forks in the Road: The Many Paths of Arts Alumni," in *Strategic National Arts Alumni Project: 2010 Findings* (Bloomington: Center for Postsecondary Research, School of Education, Indiana University, 2011), 17.

4. Future of Music Coalition, "Artist Revenue Streams" (Washington, DC: Future of Music Coalition, 2011).

5. Andrea Hausmann, "German Artists between Bohemian Idealism and Entrepreneurial Dynamics: Reflections on Cultural Entrepreneurship and the Need for Start-Up Management," *International Journal of Arts Management* 12, no. 2 (2010); Laura Hölzenspies, "Research on Professional Musician's Portfolio Careers in the Netherlands: Summary in English" (Amsterdam: Koninklijke Nederlandse Toonkunstenaars Vereniging, 2009); Royal College of Music, "Working in Music," http://www.legacyweb.rcm.ac.uk/default.aspx?pg=6823 [access date June 2014]; Paula Karhunen, "Does arts training produce too many artists? Considerations on professional training and employment in the arts," paper presented at the Third Nordic Conference on Cultural Policy Research, Bø in Telemark, Norway, August 23–24, 2007.

6. Hölzenspies, "Research on Professional Musician's Portfolio Careers."

7. Royal College of Music, "Working in Music."

8. See http://www.ae2n.net.

9. See http://usasbe.org/interestgroups/arts/.

10. See www.music.org.

11. www.ae2n.net.

12. See www.artivate.org.

13. Gary D. Beckman, ed., *Disciplining the Arts: Teaching Entrepreneurship in Context* (Lanham, MD: Rowman and Littlefield Education, 2011).

14. Committee on Music Entrepreneurship Education, *2010 CMS Summit Handbook* (Missoula, MT: College Music Society, 2010).

15. Angela Myles Beeching, "Career Development and Entrepreneurship across the Curriculum: Best Practices in Professional Development Programs in Undergraduate Music Programs," in *The Musician in Creative and Educational Spaces of the 21st Century*, ed. Michael Hannan and Lindsey R. Williams (Shanghai: Shanghai Conservatory of Music, International Society for Music Education Commission for the Education of the Professional Musician, 2010). Committee on Music Entrepreneurship Education, *2010 CMS Summit Handbook*. Kaija Huhtanen, "Towards Creative Entrepreneurship," in *The Musician in Creative and Educational Spaces of the 21st Century*, ed. Michael Hannan and Lindsey R. Williams (Shanghai: Shanghai Conservatory of Music, International Society for Music Education Commission for the Education of the Professional Musician, 2010).

16. See www.musiccareernetwork.org.

17. "Undergraduate Learning/Teaching Recommendations for Career Development in Music," *Network of Music Career Development Officers*, 2011, www.musiccareernetwork.org [access date June 2014].

18. Angela Myles Beeching, *Beyond Talent: Creating a Successful Career in Music*, 2nd ed. (Oxford: Oxford University Press, 2010).

19. David Cutler, *The Savvy Musician: Building a Career, Earning a Living and Making a Difference* (Pittsburgh, PA: Helius Press, 2010).

20. The schools are Boston University, New York University, and Syracuse University. All other programs involve direct enrollment at a foreign institution or are organized by third-party providers. See Patrick M. Jones, "Developing Intercultural Understanding through Study Abroad Programs for Music Majors: An Analysis and Critique of Current Practice," unpublished paper delivered at the conference "Cultural Inclusiveness and Transparency of Purpose in Contemporary Music Education: The Influence of Cultural, Educational, and Media Policies," Athens, Greece, July 10–13, 2012.

21. C. Robert Mesle, *Process-Relational Philosophy: An Introduction to Alfred North Whitehead* (West Conshohocken, PA: Templeton Foundation Press, 2008). 3–10.

Chapter 16

Inclusive Cultural Policy in the Music Education of Nanjing and Hong Kong

Rita Lai Chi Yip and Ji Hong Ye

In recent years, fast-growing digital technology, transportation systems, and wireless communication networks have dramatically increased contact between people from different countries and regions. People can meet each other much more easily virtually and in reality. Accompanying this is the need for better understanding among different cultures to avoid conflicts and to attain harmony in coexistence in this global village. As an important part of culture, music facilitates the multifaceted understanding of people, places, and practices. Constructing multicultural perspectives in music education is not uncommon. In this paper, the focus is on culturally inclusive policy in the music education of Nanjing and Hong Kong and on how cultural understanding can be brought about by the music curriculum. The paper analyzes the depth and breadth of this cultural understanding.

CULTURAL INCLUSIVENESS IN MUSIC EDUCATION

The inclusion of music from different parts of the world in music education may or may not be decreed by government. One of the nine National Standards for Music Education followed by schools in the United States is "understanding music in relation to history and culture" (MENC, 1994). It is not exactly government policy, but the standards are well-known to music educators in and outside the United States. American music educator Patricia Shehan Campbell has long been famous for advocating world music (Campbell, 1991, 1996, 1998, 2004). In recognizing that music is inseparable from many traditional cultural practices, Dunbar-Hall (2005) has promoted music curriculum as cultural studies, which requires teachers to be aware of the social and political roles of music in various cultures. He questions whose music should be studied in classrooms and who has the right to make such decisions. He disapproves of Eurocentric ways of analyzing non-Western music events and artifacts and of prioritizing Western ways of studying culture. Swanwick (1999) understands that popular music competes with conventional curricula in schools. He questions the need for a school music curriculum, noting that the music taught in a classroom context is not as authentic. As various forms of music culture and subculture make their way into the curriculum in many parts of the world, whether or not they are supported by government policy, an investigation into the music curricula in Chinese regions is timely. The investigation will show more clearly the extent to which culturally inclusive policy is reflected in music curricula and will inform educators and policy makers about implications for music education, teacher education, and policy setting.

METHODOLOGY

The methodology employed in this research stemmed from the comparative education field. Qualitative as well as quantitative method was employed. The curriculum documents have been analyzed qualitatively from the perspectives of an indigenous insider and an indigenous outsider (Acker, 2001). The indigenous insider is someone who is trained in and studies his or her own field or discipline. The indigenous outsider is "someone who belongs to the category yet takes a different view than those fully encapsulated within the category and for that reason is seen by the community to be at least a partial outsider. The border might be a good vantage point for a critical perspective" (Acker, 2001: section 5). Both authors of this paper are engaged in music-teacher education but in two distant regions of greater China, Nanjing and Hong Kong. They took on the roles of indigenous insider and indigenous outsider (see Table 16.1) in reviewing the curriculum documents issued and followed in their own region and the region of the other researcher. Research by an indigenous insider ensured a fuller interpretation of the conditions. Research by an indigenous outsider elicited more objective and critical views.

Table 16.1 Qualitative research method in analyzing the curriculum

	Curriculum (Nanjing)	Curriculum (Hong Kong)
Researcher A	Indigenous insider	Indigenous outsider
Researcher B	Indigenous outsider	Indigenous insider

We analyzed the curriculum documents, the intended curriculum, and the enacted curriculum (Hume and Coll, 2010; Kurz et al., 2010). The intended curriculum includes official curriculum documents issued by the government, local or national. These documents were *Music Curriculum Guidelines (Experimental Version)* (MoE, PRC, 2001) and *Arts Education Key Learning Area: Music Curriculum Guide (Primary 1–Secondary 3)* (CDC, 2003). The former is followed by schools in Nanjing, and the latter by schools in Hong Kong. The intended curriculum documents were analyzed to see how music from different cultures is laid out and what learning is expected. The enacted curriculum includes two sets of music textbooks approved by the government for use in Nanjing and Hong Kong primary schools, respectively: *Phoenix National Standard Teaching Materials: Music* (Xu and Dai, 2002) and *New Territories in Music* (Cham-lai, 2006). These enacted curricula are for implementation by teachers and are actually used by students (Hume and Coll, 2010). With the understanding that "(a) The intended curriculum reveals the instructional content targets for the enacted curriculum (i.e., what content ought to be covered in the classroom), (b) the enacted curriculum highlights the content that students had the opportunity to learn" (Kurz et al., 2010: 132), the analysis of both helped provide a comprehensive view of the issues under investigation. The two sets of music textbooks each consists of twelve volumes for primary grades one through six (with two volumes for the two semesters in each level of study). Content materials selected from non-Chinese origins for singing and listening were analyzed quantitatively to see how far culturally inclusive policy is in place. The non-Chinese content materials were further divided into Western classics and nonclassics (including folk and pop music) for quantification to gain a clearer picture of the breadth of culturally inclusive policy implementation in school music education in these regions.

CURRICULUM REFORM

The music curriculum followed by schools in Nanjing and Hong Kong has undergone tremendous changes in the twenty-first century. Since the start of the century in mainland China, *Music Curriculum Standards (Experimental Version)* for primary and junior secondary schools has been issued by the Ministry of Education of the People's Republic of China (MoE, PRC, 2001). Then, in 2003, *Senior High Music Curriculum Standards (Experimental Version)* (MoE, PRC, 2003) was issued. Under the one-country, two-system policy, Hong Kong had its new *Music Curriculum Guide (Primary 1–Secondary 3)* in 2003 (CDC, 2003); in 2007 the new *Music Curriculum and Assessment Guide (Secondary 4–6)* (CDC and HKEAA, 2007) came into being. The chief reason behind the curricula reform in mainland China was apprehension that "the older educational perspectives, contents, pedagogies, tactics and assessment means cannot cope with the demands of the development of a quality education" (MoE, PRC, 2001: 1). Besides, it was put forth that "world peace and the development of the world depends on understanding and respect for different cultures." Thus the foundation for inclusion of a range of music from different parts of the world is well set.For Hong Kong, the rationale for opening up the curriculum for culturally inclusive policy is clear: "To meet future challenges, schools are shouldering the important mission to cultivate in students broader perspectives, creative thinking, rich knowledge, flexibility and a strong sense of commitment. On such a firm basis, students can engage in life-long learning and incessantly improve themselves so as to meet future challenges" (CDC, 2003: 3).It can be seen that changes in the music curriculum of both Nanjing and Hong Kong are similar in the trend of including more contents for the enrichment and broadening of students' perspectives. This is a pretext for a culturally inclusive policy in the music curriculum.

CULTURAL INCLUSIVENESS AND THE INTENDED MUSIC CURRICULUM OF NANJING

The curriculum document followed in Nanjing indicates that it values music as an important carrier of human cultural heritage and as a demonstration of human wisdom. The message is that music as culture is an angle of learning (MoE, PRC, 2001: section 4) and that "through learning the music culture of other countries and nationalities, students will broaden their aesthetic vision, learn the richness and diversity of the musical culture of different nationalities, and enhance understanding, respect and love of different cultures" (MoE, PRC, 2001: section 4). The breadth of learning extends to different countries and nationalities. The focus is on the aesthetic dimension and diversity and on learning, respecting, and loving different kinds of music. The focus is also to teach students the value of equity and to facilitate the sharing of human civilization and its distinct outcomes (MoE, PRC, 2001: section 9).On the affective side, the ultimate aims are the nurturing of love for relatives, humankind, and all good things, as well as development of a positive and optimistic attitude toward life and the desire for a better future (MoE, PRC, 2001: part 2). Diverse ethnic styles and emotions embedded in the music are to be learned, and students are to develop a love for the music of other peoples in the world.The curriculum document provides guidance in encouraging students to listen to music and to understand the close ties between music and life. There are content standards on musical styles and schools of music, children's songs, nursery rhymes, music for small instrumental ensembles, and music clips from different countries, regions, and ethnic groups so that students can learn about different styles.In addition to listening to different world ethnic folk music, students have to comment on its style and characteristics. They also listen to the best music of the world to understand different musical genres. The study of music can directly enhance students' cultural literacy and expand their cultural horizons. The curriculum document also points out the relationship between music and nonarts disciplines and the relationship between sound art and language art. The document includes representative children songs, nursery rhymes, fairy tales, and poetry, as well as music from certain historical periods, regions, and countries. The document also suggests the application of social science and natural science to deepen an understanding of musical works and of customs of other countries as presented by songs and music. Overall, it is suggested that music and related culture can be truly understood and experienced only through music appreciation, performance, and creative activities. Particulars about the use of information technology in music teaching are deliberated. The combined use of sound and images via modern information technology is encouraged.From the indigenous outsider's viewpoint, the curriculum standards followed by Nanjing schools provide many detailed guidelines concerning the inclusion of music from the diversity of world culture. Even today, this 2001 document is relevant in terms of a combination of timely, fundamental, and selective aspects concerning cultural inclusivity in the curriculum.

CULTURAL INCLUSIVENESS AND THE INTENDED MUSIC CURRICULUM OF HONG KONG

The curriculum followed in Hong Kong is less detailed than that followed by Nanjing, although the book is thicker. The focus of the Hong Kong music curriculum regarding cultural inclusiveness is more on the value of music and music appreciation: "understanding and respecting traditions as well as values of other cultures through appraising music from different cultures" (CDC, 2003: 15). The aim is to cultivate critical responses to music, beginning with describing and then analyzing music. Similar to the curriculum followed by Nanjing, it discusses the relationship between words and music. Examples include "word painting, syllables in English and tones of Cantonese dialect" (CDC, 2003: 28).Understanding music in context is pointed out. The goal is to describe music of different styles and cultures in context. Examples include:

- Discussing the effects of nineteenth-century nationalism on Western music

- Discussing how impressionism in visual arts affected the music of Debussy

- Discussing the role of music and the use of sound effects in films

- Discussing how social values influence the style and content of popular songs in different historical contexts

- Researching and discussing the cultural and historical contexts of a particular type of music (CDC, 2003: 30)The curriculum advises teachers to use quality and suitable songs to stimulate student interest and enhance ability: "Teachers should choose songs with various levels of difficulty, from different cultures and styles so as to broaden students' music horizon and foster their interest in singing" (CDC, 2003: 54), and lessons should "provide quality music repertoire in different cultures and styles" (CDC, 2003: 83). One activity is to "identify the source of a piece of music such as its originating country and culture" (CDC, 2003: 76). The guidelines include materials on musical instruments, composers, and their music (CDC, 2003: 160–61).

- General Comparison of the Enacted Curriculum in Nanjing and Hong Kong

In general there is a gap between the amount of non-Chinese music materials for singing and listening in Nanjing and Hong Kong primary schools. The numbers for Nanjing primary schools are smaller (ranging from 1 to 38 percent of GDP (see Figure16.1); those for Hong Kong primary schools range from 52 to 72 percent. Coincidently, the figure for grade P6 for both Nanjing and Hong Kong is 53 percent. On average, non-Chinese music materials for singing and listening in Nanjing and Hong Kong primary schools amount to 31 and 65 percent, respectively. The difference might be accounted for by the history and politics of Hong Kong being under significant Western influence.

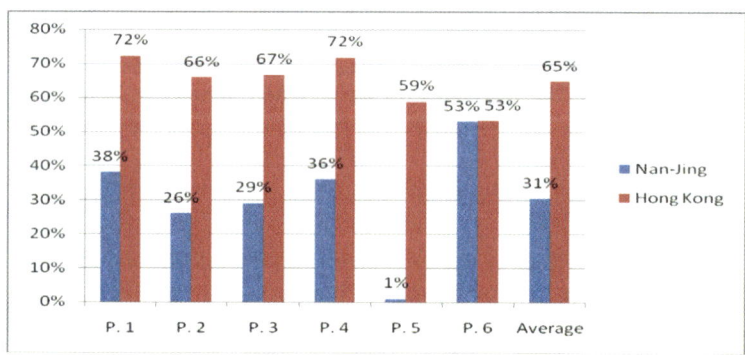

Figure 16.1 Comparison of non-Chinese music materials for teaching in primary schools

THE ENACTED CURRICULUM IN NANJING

As a whole, the non-Chinese singing materials for grades P1 through P6 are chiefly world music (see Figure 16.2), meaning folk songs from different parts of the world, film music, and songs for special occasions, such as Christmas. Although for grade P6 there is an exception, with 10 percent of singing materials from the Western canon, this is not included in the figure, since it is the only class with these materials. In this case, the materials include mostly works from the Classical and Romantic periods. Works from the twentieth century are very few; pop music is also minimal. The indigenous insider researcher explains that music teachers mostly rely on their studies at universities, which focus chiefly on the Western Classical and Romantic periods. Music teachers are relatively unfamiliar with the masterpieces of modern music and pop music.

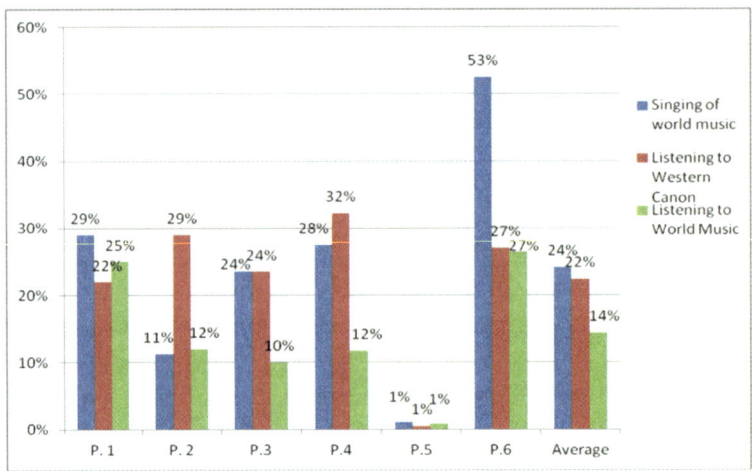

Figure 16.2 Non-Chinese classic or folk music enacted for each level of primary class (Nanjing)

On average, world music singing materials (24 percent) are more prevalent than listening materials (14 percent). This is especially true for grade P6 (with singing materials accounting for 53 percent) and P1 (29 percent). Students in these two grades have comparatively more chances to listen to world music (27 percent for P6; 25 percent for P1). For grades P2, P3, and P4, listening materials are more on the Western canon.

THE ENACTED CURRICULUM IN HONG KONG

The enacted curriculum in Hong Kong analyzed included the singing of and listening to Western canon and world music. Although the category of Western canon singing is included, the percentage is very low (4 percent) (see Figure 16.3). The singing of world music is over 50 percent for grades P2, P4, and P6, with 62 percent, 68 percent, and 50 percent, respectively. The songs may be in foreign or Chinese lyrics. Foreign lyrics may be in English or occasionaly in phonetic versions of languages such as Japanese, Filipino, or Indonesian. The numbers for listening to the Western canon are much higher: 54 percent on average and as high as 86 percent in P1; percentages in P3 and P4 are also over 50 percent (66 and 64 percent, respectively). The generally high percentage of listening to the Western canon is no surprise in the view of the indigenous insider researcher, as this reflects the highly Westernized music education policy in Hong Kong.

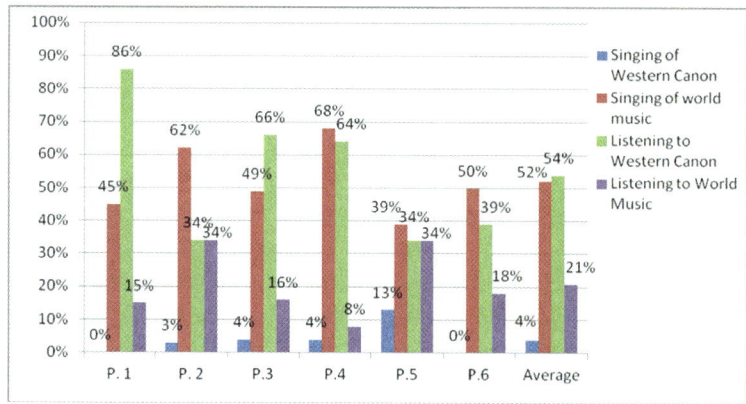

Figure 16.3 Non-Chinese classic or folk music enacted for each level of primary class (Hong Kong)

Although the percentage of singing materials of a world music nature is high and quite similar to the percentage of listening to the Western canon (52 and 54 percent, respectively), the percentage of listening to world music is much lower (21 percent). The phenomenon is similar to that of the enacted curriculum in Nanjing and might be due to the limited training of music teachers in world music, especially in the pure instrumental music domain.

TOWARD A CULTURALLY INCLUSIVE POLICY IN MUSIC EDUCATION

In reviewing the intended music curriculum and the enacted music curriculum of Nanjing and Hong Kong, we find that awareness of culturally inclusive policy is evident. The degree of awareness might be different due to historical and political reasons. This in turn impacts music-teacher education programs. The Western canon—specifically works of a more tonal nature—occupies a prominent position in the enacted curriculum followed by Nanjing. World music singing materials account for comparatively higher percentages than do more authentic instrumental listening materials in both Nanjing and Hong Kong.

From the perspectives of both the indigenous insider and indigenous outsider researchers, the Nanjing music curriculum, whether intended or enacted, has much more respect, openness, tolerance, and understanding regarding different cultures of the world in the selection of materials and course layouts. There is innovation in philosophy, teaching methods, teaching strategies, and teaching content. School music education has opened to diverse world cultural horizons. The concept of nurturing quality personnel with contemporary cultural perspectives has deepened through various experimental music classes, extracurricular music activities, and art practices. Limitation in the further implementation of culturally inclusive policy in school music education may be checked when similar policy can be enforced in music-teacher education. With sufficient professional development for preservice and in-service music teachers, there could be more advancement in implementation of the policy. As for the Hong Kong music curriculum, from the perspective of the indigenous outsider researcher, it appears to have far more cultural inclusiveness, with more modern/contemporary music, pop music, and other innovative sound initiatives, especially in the enacted curriculum. It is also assumed that there is less restriction in extending the curriculum into new territories. Although the view regarding less restriction is shared, from the perspective of the indigenous insider, the situation has not improved compared to that of ten years ago. Specifically, there were higher percentages of foreign songs (from 66.3 to 77.8 percent), whether with foreign or Chinese lyrics, in textbook series (Lai and Yip, 2000). Despite regression on the part of Hong Kong, the principle of openness in the designing of music curriculum standards is still visible in both Nanjing and Hong Kong. Both regions are cautious about correctly handling the relationship between traditional and modern music, classic and nonclassic music, and Chinese musical culture and the world's diverse culture. Music teachers are urged to strive for a balance in using outstanding works from different cultures, with a modern social sensibility, to enrich teaching materials and to broaden the musical horizons of students.

CONCLUDING OBSERVATIONS

In a world where pluralistic societies are increasing, multicultural education has gradually become the norm. A successful multicultural education depends on a feasible culturally inclusive policy. Since cultural inclusiveness in music is comparatively easier to implement than in other subjects or disciplines, a concrete culturally inclusive policy in music education could have increased impact on the establishment of equity and diversity in the world. A successful culturally inclusive policy in music education—whether in Nanjing, Hong Kong, or elsewhere—would require music educators to learn and unlearn. It is a journey to better the intended and enacted curriculums.

Note: This chapter is reprinted with permission of Professor J. Scott Goble and Professor Tadahiko Imada. An earlier version of this chapter appeared in *Cultural Inclusiveness and Transparency of Purpose in Contemporary Music Education: The Influence of Cultural, Educational, and Media Policies* (University of British Columbia, 2012).

BIBLIOGRAPHY

Acker, Sandra. 2001. "In/out/side: Positioning the Researcher in Feminist Qualitative Research." *Resources for Feminist Research* 28, no. 3–4: 153–72.

Campbell, Patricia Shehan. 1991. *Lessons from the World: A Cross-Cultural Guide to Music Teaching and Learning*. New York: Schirmer Books, 1991.

———. 1996. *Music in Cultural Context: Eight Views on World Music Education*. Reston, VA: Music Educators National Conference.

———. 1998. *Music Resources for Multicultural Perspectives*. Reston, VA: Music Educators National Conference.

———. 2004. *Teaching Music Globally: Experiencing Music, Expressing Culture*. New York: Oxford University Press.

Cham-Lai, Suk Ching. 2006. *New Territories in Music*. Hong Kong: Education Publisher Limited Company.

Curriculum Development Council (CDC). 2003. *Arts Education Key Learning Area: Music Curriculum Guide (Primary 1–Secondary 3)*. Hong Kong: Curriculum Development Committee, http://www.edb.gov.hk/index.aspx?nodeID=2689&langno=1 (accessed December 30, 2006).

Curriculum Development Council and the Hong Kong Examinations and Assessment Authority (CDC and HKEAA). 2007. *Music Curriculum and Assessment Guide (Secondary 4-6)*, Hong Kong: Education and Manpower Bureau, 2007, http://www.edb.gov.hk/index.aspx?nodeID=2689&langno=2 (accessed January 30, 2008).

Dunbar-Hall, Peter. 2005. "Colliding Perspectives? Music Curriculum as Cultural Studies." *Music Educators Journal* 91, no. 4: 33–37.

Hume, Anne, and Richard Coll. 2010. "Authentic Student Inquiry: The Mismatch between the Intended Curriculum and the Student-Experienced Curriculum." *Research in Science and Technological Education* 28, no. 1: 43–62.

Kurz, Alexander, Stephen N. Elliott, Joseph H. Wehby, and John L. Smithson. 2010. "Alignment of the Intended, Planned, and Enacted Curriculum in General and Special Education and Its Relation to Student Achievement" *Journal of Special Education* 44, no. 3: 131–45.

Lai, May Tan, and Lai Chi Rita Yip. 2000. "Content Analysis of Selected Primary Music Textbook Series." In *School Curriculum Change and Development in Hong Kong*, edited by Yin Cheong Cheng, King Wai Chow, and Kwok Tung Tsui, pp. 411–24. Hong Kong: Hong Kong Institute of Education.

Ministry of Education, People's Republic of China (MoE, PRC). 2001. *Music Curriculum Guidelines (Experimental Version)*. Beijing: Beijing Normal University Publisher, http://ywjy.cersp.com/kbyj/kcbz/200511/349.html (accessed February 19, 2011).

———. 2003. *Senior High School Music Curriculum Guidelines (Experimental Version)*. Beijing: Beijing Normal University Publisher.

Music Educators National Conference (MENC). 1994. *The School Music Program: A New Vision,* http://www.menc.org/resources/view/the-school-music-program-a-new-vision (accessed January 6, 2007).

Swanwick, Keith. 1999. "Music Education: Closed or Open?" *Journal of Aesthetic Education* 33–34: 127–41.

Xu Zoya, and Dai Haiyun. 2002. *Phoenix National Standard Teaching Materials: Music.* 12 vols. Nanjing: Jiangsu Children's Publishing House.

Yip, L. C. R., and J. H. Ye. 2012. Cultural Inclusive Policy in the Music Education of Nanjing and Hong Kong. 2012. In *Cultural Inclusiveness and Transparency of Purpose in Contemporary Music Education: The Influence of Cultural, Educational, and Media Policies,* edited by J. S. Goble and T. Imada, Vancouver, BC: University of British Columbia.

Chapter 17

Reshaping the Creative Culture of Hong Kong Cinema under Chinese Shadows: Marketing Coproduction Films in Mainland China and the Preservation of Localism

Patrick Yeuk-Kwong Yuen

The growth of the film market of mainland China in the last decade has brought significant changes to current Hong Kong cinema. Through the implementation of an economic agreement, the film industry of Hong Kong is now able to engage the emerging market on the mainland. The circumstances drive Hong Kong cinema closer to a restructuring of the national film market of China.

Under the political context of "one country, two systems," Hong Kong cinema in general retains its pre-1997-handover heritage, characterized by liberalism and commercialism. The mainland and Hong Kong Closer Economic Partnership Agreement (CEPA), signed in 2003, has provided an opportunity for the film industries of both areas to collaborate. Aiming to enhance trade and investment between the mainland and the Hong Kong Special Administrative Region, CEPA includes commitments that allow, with certain requirements, Chinese-language motion pictures, jointly produced motion pictures, and cinema theater services of Hong Kong moving to the mainland.[1] Cantonese films from Hong Kong, as approved by the State Administration of Radio, Film and Television (SARFT),[2] can be shown in Guangdong Province if introduced by the provincial government from 2007. The latest supplement of the agreement allows dialectic versions, in Cantonese, to be shown on the mainland if properly subtitled and released through a centralized distribution service. The agreement helps investors, producers, and filmmakers find paths to a vast market. Hence the film industry of Hong Kong, after the recession of the mid-1990s, underwent major reshuffling, which substantially affected the creative culture of filmmaking.

A crossroad lies before many filmmakers in Hong Kong. One option is to gain access to a prosperous new market on the mainland with a population of 1.3 billion. This could mean ten times the box office revenues. But the opportunity comes with the challenges of meeting expectations of a vast audience with a socioeconomic and cultural background very different from that of Hong Kong. Further hurdles are the restraints of censorship from Chinese authorities. Chinese film censors are known to be sensitive to political and ideological issues embedded in content.

Another option is to hold onto the interests of audiences in Hong Kong while accepting the limitations of scarce financing and risky revenues from a shrinking local market. The heritage of creative liberty in Hong Kong allows unrestricted exploration of film genres and artistic choices for theme or content. However, the limited local box office potential leaves low incentive for investors. As a result, financing productions, especially those calling for big budgets, has become difficult.

This article examines the current trend of coproduction films with the mainland and how filmmaking and industrial conditions are affected in Hong Kong. It also looks into the subsequent changing creative culture of Hong Kong cinema and the challenges, as well as opportunities, as it integrates itself into China's national film industry.

REVIVING THE FILM INDUSTRY OF HONG KONG THROUGH COPRODUCTION

The film industry in Hong Kong had endured a continuous decline since the mid-1990s. Nearly two hundred films a year were produced for commercial distribution in Hong Kong in the 1980s, which is considered the golden age of the industry. However, 126 films were produced in 2001, and the number was down to 52 in 2006. The total box office revenue generated from locally produced films in 2006 was about HK$253 million (US$32.4 million), almost half of more than HK$547 million (US$70.1 million) in 1997.[3] Chains of movie theaters showing locally produced films dropped out of prosperous downtown areas, giving way to multiplex cinemas showing mostly Hollywood movies. Their withdrawal from the market drained funds for new productions. Many attributed the setback to the stagnant quality of movies produced locally. Keen competition from Hollywood blockbusters, enriched by costly computer-generated images and sophisticated high-concept drama, captured the middle-class audience. Some said that the diversity of entertainment disrupted the habits of moviegoers. Copyright infringement in form of video piracy was another problem. In addition, subsiding sales of Hong Kong films to Taiwan, South Korea, and Japan discouraged investors from financing local film projects.

To many optimists, access to the emerging market on the mainland through coproduction brings hope of a revival of the film industry in Hong Kong. As of 2004, per CEPA, motion pictures jointly produced by Hong Kong and mainland film companies are treated as motion pictures produced on the mainland. They are exempted from the quota system, an entry hurdle for all foreign films seeking import to the mainland. For example, only twenty foreign films were allowed to enter the mainland in 2001, the year China gained accession to the World Trade Organization. The annual quota from the United States has recently been increased to thirty-four, including fourteen films of advanced technological format released in 3D or IMAX, according to an agreement memorandum signed between the two countries in 2012.

Making coproduction films signals a long-awaited lucrative alternative for film producers in Hong Kong, who feel it is risky to rely solely on local box office revenues. The pursuance of the mainland market appears to be paying off and it continues to develop. More than thirty coproduction films were released in 2010, a 50 percent increase from 2005. The revenue from theatrical release of coproduction films was more than RMB 2.3 billion (US$374.8 million) in 2009, compared to RMB 579 million (US$94.1 million) in 2005.[4]

Table 17.1: Box office revenue of coproduced films in mainland China, 2005–2010

Year	2005	2006	2007	2008	2009	2010
Number of films	20	23	23	29	29	30+
Box office revenues	5.79	8.14	8.65	20.27	23.61	-

Box Office Revenue figures in RMB $100 million

RESTRUCTURING THE FILM MARKET IN CHINA: FROM NEAR COLLAPSING TO BOOMING

The film industry in China was under strict state control for decades. The traditional communist conception of film as an important tool of propaganda excluded the commercial mind-set regarding this mass medium. The content of films was vetted by a hierarchal censorship system. Studios and theaters run by state-owned enterprises followed a rigid economic plan masterminding film production and distribution. As Graeme Turner observed, the regulation and control of the national film industry was of great importance because it meant restriction and limitation of representations of the nation, which "produce and reproduce the dominant point of view."[5] Even after economic reform started in early 1980s, concerns of ideological influence prevented the film industry from reforming itself as other sectors did.

Under an old-fashioned communist film policy, strict censorship controlled the creative operation of mainland Chinese filmmakers, who were relieved from the burden of achieving box office results but in return were discouraged by tedious ideological vetting procedures. State-owned studios made movies of "main melody"[6]—loaded with didactic or propaganda content—to meet requirements of governmental policy. However, audiences nationwide lacked interest in such movies. On top of scanty admissions, mandatory low ticket prices inflicted further income deficits. Film studios ran low of initiatives to produce films unless assigned to do so. Declining movie attendance, a lack of funding for production, and a low supply of films for release caused serious setbacks to the film market in China. Attendance dropped from 14.39 billion in 1991 to 10.55 billion in 1992. Box office revenues were RMB 1.99 billion in 1992, a nosedive of more than 20 percent from 1991. Six out of the sixteen film studios in the country faced the threat of bankruptcy.[7]

The threat of a collapsing film industry urged the central government to consider adapting a change of policy. Under economic reform commenced in the 1980s, the long-awaited marketization of the film industry in China began to take place under cautious control. Starting in 1993, film studios were allowed to deal directly with domestic distributors in different provinces and cities. This led to the end of the monopoly of the state-owned China Film Distribution and Exhibition Company. To activate the film market, a limited number of foreign films, mostly from Hollywood, were introduced in 1995. Two years later, production companies with funds from the private sector were allowed to enter the business. As Xiaoli Li suggests, private companies like Beijing Polybona Films and Huayi Brothers Media grew into resourceful business organizations. Together with the powerful state-run China Film Group and Huaxia Film, a joint venture of formerly state-run film studios, they currently play significant roles in investment, production, and distribution of films in China, especially blockbuster movies.[8] A prototype of a market economy was created for the Chinese film industry right before China's accession to the World Trade Organization in 2001.

Since 2005, box office revenues in China have grown at a two-digit rate almost every year.[9] (See Table 17.2.) Surpassing Japan in 2012, China became the second-largest film market in the world, with domestic theatrical revenue of RMB 17 billion (US$2.7 billion), second only to the North American market merging the United States and Canada.[10] The Chinese box office saw a record growth of 36 percent in 2012.[11] The Chinese box office revenue in 2013 reached RMB 21.77 billion (US$3.58 billion).[12] EntGroup Inc., a research firm specializing in China's entertainment business, attributes the prosperous development of the mainland film market in recent years to the following: an increasing number of cinemas and screens in China; an emerging middle class that has embraced moviegoing; and significant growth in film production, distribution, and financing.[13] According to a report released by the China Communication Center of SARFT in July 2013, there were 13,118 movie screens in China at the end of 2012. The figure represented an annual growth of 3,832 screens, with an average of 10.5 new screens per day,[14] a huge leap compared with the approximately 9,200 screens in 2011 nationwide and a growth rate of 8.3 screens per day.[15] (See Table 17.2.)

Table 17.2 Box office revenue of mainland China, 2005–2012

Year	2005	2006	2007	2008	2009	2010	2011	2012
Box office revenue	20.46	26.20	33.27	43.41	62.06	102.52	131.15	170

The substantial growth of the film market on the mainland impresses investors—domestic and foreign alike—and the coproduction of films plays a major part in the current overall booming picture. The film industry of Hong Kong plays a significant part in the growth. Seven out of the top ten box office hits in 2012 in the domestic production category were coproduction films with Hong Kong producers and filmmakers.[16] (See Table 17.3).

Table 17.3 Top ten in box office revenue, 2012—domestic productions

Rank	Film Titles	Box Office Revenue (RMB million)	Coproduction with Hong Kong
1	Lost in Thailand (人再囧途之泰囧)	116.98*	
2	Painted Skin II: The Resurrection (畫皮II)	70.45	✓
3	Chinese Zodiac (十二生肖)	73.59*	✓
4	Back to 1942 (一九四二)	37.20	
5	Cold War (寒戰)	25.36	✓
6	The Silent War (聽風者)	23.37	✓
7	The Four (四大名捕)	19.21	✓
8	The Great Magician (大魔術師)	17.41	✓
9	Caught in the Web (搜索)	17.35	
10	Mission Incredible: Adventures on the Dragon's Trail (喜羊羊與灰太狼之開心闖龍年)	16.59	✓

*: Revenue until January 6, 2013

BALANCING COMPETITION IN A GLOBALIZING FILM MARKET

Globalization plays an important role in integrating the world into a unified market for international distribution of films. The doctrine of free trade has opened many doors to markets where access had been blocked since the cold war due to ideological differences. The box office of countries in the Asia Pacific region saw 15 percent growth in 2012, and China experienced the greatest leap of 36 percent.[17] However, the film market plateaued in most developed countries, such as western European nations and the United States. The box office in the United States–Canada grew 6 percent to US$10.8 billion in 2012, which marked the first rise in three years. Even though revenue figures have remained steady and even increased slightly, general ticket sales have been declining for ten years in North America.[18] The lowest number of tickets sold since 1995 was recorded in 2011, with 1.29 billion, compared to 1.6 billion tickets sold in 2002. The total number of tickets sold in 2012 was 1.36 billion, which represented a rise of 5.6 percent from the previous year.

For more than a decade, Hollywood has been counting on foreign sales instead of revenues from domestic theatrical releases. For example, *The Avengers,* a sci-fi action adventure released in 2012 by Disney, brought in US$623 million from the domestic market while US$1.5 billion was collected from overseas markets.[19] The film generated US$93.9 million (RMB 575.95 million) in the mainland market. It was ranked twentieth-third among top box office hits in China, including domestic and imported foreign films.[20]

The quota system for films imported to China keeps the influx of foreign films manageable. However, it does little to improve the lack of competitiveness of many domestic films and their performance at the box office. The biggest overall box office hit in China, as of early June 2013, was the legendary sci-fi action adventure *Avatar,* which set a record of US$226.8 million (RMB 1.39 billion) in revenue in 2010. It was also the all-time highest-grossing movie in the world, with global revenue of US$2.78 billion (RMB 17.07 billion).[21] A Chinese comedy released in 2012, *Lost in Thailand,* held second place after *Avatar* with revenue close to US$206.6 million (RMB 12.6 billion). It aroused heated debate that imported Hollywood films, although limited to twenty per year,[22] had taken the bulk of box office revenue after they began entering China. In 2012 there were 893 domestic film productions, including feature and documentary films. About one-third of these films were released in movie theaters nationwide, and the total gross was RMB 8.27 billion. However, the figure represented less than half of the market share; the rest was taken by eighty-six imported films.[23] Some Chinese critics disapprove of the trend and plea for measures to nourish domestic film productions. Among such measures are blackout periods imposed by the authorities. During these periods, foreign imported films would not be permitted to open in theaters, protecting domestic films from head-to-head competition.[24]

OPPORTUNITY AND RESTRAINT FOR HONG KONG'S COPRODUCTION FILMS IN THE MAINLAND MARKET

Coproduction in filmmaking between mainland China and Hong Kong can be traced back to 1983, when Li Han-hsiang (李翰祥), a famous film director in Hong Kong, sought to produce a period film based on a historical event in the late Ching Dynasty. The film, *The Burning of the Imperial Palace* (火燒圓明園, 1983), with scenes shot on location at the Forbidden City in Beijing, was produced on a coproduction basis. At the time, no strict regulation was stipulated regarding the cooperation of funding or personnel from either side. Although the film turned out to be a commercial and critical success in Hong Kong, distribution and theatrical release on the mainland were restricted. Cold war ideological differences and the absence of an open film market in China kept most foreign films, especially Western productions, from mainland movie screens until the early twenty-first century.

After China's accession to the World Trade Organization in 2001, China steadily engaged the world with free trade.[25] Steps heading to a market economy entailed a policy change to transform the once-closed film market. For example, China now agrees to allow foreign participation in joint ventures in the distribution of video and sound recording products from the United States and elsewhere. Twenty films from the United States are imported to China for revenue sharing each year.[26] In addition to coproductions with Hong Kong and the United States, coproductions with other foreign countries are possible under a regulation issued in 2003 by SARFT.[27]

To filmmakers in Hong Kong, CEPA could be considered part of deregulation of the policy of importing films to the mainland. It enables the growth of coproductions between film producers and investors from the mainland and Hong Kong and sets a new agenda for filmmakers to explore a new market on the mainland. However, there are CEPA requirements to fulfill before a coproduction film can commence. Furthermore, a process of content vetting is crucial to obtain official approval from Chinese censoring authorities.

Among the requirements in CEPA, a Hong Kong–registered production company must own more than 75 percent of the copyright of the film, and Hong Kong residents must make up more than half of the principal personnel, such as the director and leading actors. To fulfill coproduction requirements, one-third of the leading artists, the plot, or the leading characters must be related to the mainland.[28] At least one-third of the investment should also come from the mainland.

CHANGING CREATIVE CULTURE IN FILMMAKING: APPEALS TO THE MAINLAND MARKET OR LOCAL AUDIENCES IN HONG KONG

Many films have been produced jointly by investors from the mainland and Hong Kong since the implementation of CEPA in 2004. Nevertheless, entry to a promising market is not without restraints. For coproduction films, the first hurdle to overcome is a censorship process imposed by governmental authorities. Many filmmakers from Hong Kong put this on the top of their agendas, because approval from the authorities is vital to commence coproduction. Without final official approval for theatrical release issued by the authorities, the completed film will not be allowed to show in any cinema on the mainland.

The regulation of film in China is set and enforced by SARFT, a governmental authority under the State Council of Central Government. Rules and administrative orders issued by the State Council[29] or by SARFT regulate investment in, production of, and distribution of all kinds of films, including domestic films and coproductions. The stipulations on film censorship, among others, apply strict controls to the content of films to be imported, exported, distributed, or exhibited in China.[30] The approval process is twofold, involving first clearance of the script to be produced and second the granting of a distribution license for the final version of the film.[31] The twofold process means that the authorities can ban a film project before it is put into production and can keep a completed film, even made with an approved script, from being released—since censors of the two stages are often different and may hold different opinions.

REGULATING CONTENT: CENSORSHIP AND ADAPTATION OF FILMMAKERS

Two issues arise under the censorship system. The first concerns conceptual and ideological issues. Some rules are difficult to interpret, making it hard to determine what will cause a violation. Censorship is usually determined by subjective interpretations of censors. The absence of a film rating system makes it more difficult to estimate how a film will be categorized and limited. For instance, section 25 (1) bans films with content related to "disclosure of national secret, endanger national security, damage the honor or interest of the nation."[32] It is very difficult to define in detail what is damaging to the "honor or interest" of the country when it comes to dramatic presentation.

The second issue regarding film censorship is low transparency of the reasoning behind decisions. Investors and producers often aren't given clear rationale for disapproval of a film, and they have to rely on information from unofficial sources or even speculation. A coproduction film produced by Hong Kong filmmakers in 2008, *Shinjuku Incident*, with a cast including Jackie Chan, tells how some immigrants from China, mostly illegal, get involved in ferocious gang conflicts in Tokyo and end up dying tragically. The film was forbidden release on the mainland, but the reasons behind the decision were not announced. Publicly, film producers said that the ban was due to the abundant violent scenes, but it was known in the industry that the film was banned because of its negative images of Chinese people who had gone to Japan.

To avoid the censor's ban, filmmakers tend to explore themes or stories about ancient China that have no or little relevance to contemporary political or controversial issues. This explains why many coproduction films are period films based on historical figures and incidents. But even avoiding modern context does not necessarily mean that a film will be safe. Problems may arise if any content is found to be analogical to contemporary political, social, or ideological issues.

Though not necessary intentional, the censorship system and the discretional approach limit story content, including theme, plot, and even dramatic details, because filmmakers are daunted by a possible ban of their script or film. Some films popular in Hong Kong, such as gangster and horror movies, will not be adopted in coproduction because censorship rules clearly ban content that promotes superstition, violence, or criminal acts.[33] Movies like the Chinese kung fu exorcist film *Mr. Vampire* (殭屍先生, 1985) and the youth gangster movie *Young and Dangerous* (古惑仔, 1985),[34] which scored successful box office records in Hong Kong, would not be considered for coproduction.

FROM ACCEPTANCE TO APPLAUSE: ENTERTAINING AUDIENCES ON THE MAINLAND

Because of the censorship system in China, the development of film genres is limited. Starting with a prosperous era in the 1980s, when many young directors in Hong Kong developed new approaches to conventional genres and creative styles, genre film development has forged a "vigorous local tradition" in cinematic as well as narrative styles.[35] Engaging in coproduction films casts a new challenge to filmmakers in Hong Kong, who now face conditions unfamiliar within the creative culture they inherited. The first condition filmmakers from Hong Kong must adapt to is a creative environment with a limited spectrum for genre exploration. For example, because of the regulations mentioned above, it would be risky to adapt stories involving ghosts, sexual subjects, or criminal acts. In addition, films that give a negative depiction of government officials or law enforcers are very likely to be banned. Stories critical of political leadership, the Communist Party, government policy, or related subjects are strictly off-limits.

Another hurdle before Hong Kong filmmakers seeking a breakthrough in coproductions involves appealing to mainland audiences. For a cultural industry, commercial success is indispensable. Owing to gaps between the different socioeconomic systems in mainland China and Hong Kong, it is difficult at times to gain a warm reception from mainland audiences. It is a long and challenging learning process for filmmakers from Hong Kong to communicate effectively with audiences on the mainland, who have different tastes in entertainment. Pioneering filmmakers from Hong Kong who have attempted to discover bridgeheads in the mainland market have experienced mixed results. They have developed different creative approaches to nourish story content that mainland audiences can endorse.

According to Professor Ni Zhen at Beijing Film Academy, three different approaches have been adapted by Hong Kong cinema in attempts to integrate itself into the mainland.[36] First is to use story content that takes advantage of bizarre locations or environments on the mainland, highlighting geographic or cultural exoticism for story background. Examples are the action-adventure swordsman films directed by Tsui Hark, like *The Legend of Zu* (蜀山, 2001), *Seven Swords* (七劍, 2005), and *Detective Dee and the Mystery of the Phantom Flame* (狄仁傑之通天帝國, 2011). A musical romance directed by Peter Ho-Sun Chan, *Perhaps Love* (如果, 2005), set in a nostalgic Shanghai, is another example.

The second approach, as Ni observes, is to depict human stories on the mainland in a historical or contemporary context but from the sociohistorical perspective of Hong Kong filmmakers. The films touch on the mentality of individuals or human drama in a certain period of China's history with an extension of universal sensibility in trying to overcome cultural boundaries. *Everlasting Regret* (長恨歌, 2005), directed by Stanley Kwan, and *The Postmodern Life of My Aunt* (姨媽的後現代生活, 2007), directed by Ann Hui, are in this category.

The third approach, according to Ni, is to hold onto the "native" filmmaking characteristics of Hong Kong cinema. Examples include films like *Internal Affairs* (無間道, 2002), codirected by Andrew Lau and Alan

Mak, which succeeded at both the box office and in critical reviews. Ni is optimistic that by preserving advantages in its international vision, cultural heterogeneity, and commercial sensitivity, filmmaking of Hong Kong can continue to develop. He also suggests that while exploring the market on the mainland is an important business strategy, filmmakers of Hong Kong should not turn away from Hong Kong filmmaking's traditional position rooted in cultural openness and an international perspective.

HOLDING ONTO THE HERITAGE OF HONG KONG CINEMA: LIBERAL MIND-SET AMID RESTRAINTS

Engaging the mainland film market could be considered a process toward regionalization in the period after handover in 1997. Another tendency of filmmakers in Hong Kong is to refresh their creativity from a new localism.[37] Their efforts represent a new creative culture in filmmaking by the pursuit of low- to medium-budget projects and the exploration of subjects based on social issues or local interests, which are familiar to audiences in Hong Kong. Examples include Fruit Chan's social drama on juvenile urban alienation, *Made in Hong Kong* (香港製造, 1997), and his film depicting post-1997 social impacts, *Hollywood Hong Kong* (香港有個荷里活, 2002). The former, recognized as a reputable independent film, was said to cost only half a million Hong Kong dollars and was shot with leftover film stock. Another story was told in *Gallants* (打擂台, 2010), directed by Derek Kwok and Clement Cheng, which depicted underdog heroism enriched with reminiscent references to Hong Kong's premetropolitan era in the 1970s. Despite positive reviews and publicity drawn by the stardom of Andy Lau, the investor of the film, *Gallants* received about HK$4.5 million in Hong Kong and only RMB 1 million on the mainland, which represented a deficit against the movie's cost of HK$8 million.

These movies sought new territory beyond conventional film genres in Hong Kong cinema, but commercially they attempted to remain on the safe side, with low budgets and affordable casts. A cinematic stylization close to the Italian neorealist tradition could be found in their real location setups and improvised acting by nonmainstream and amateur performers. Films based on localism like these were acknowledged with favorable reviews, but few scored successful box office records. Their objective to break even in revenue somehow highlights a workable approach to keep films of local interest feasible.

The mass-audience appeal of exotic or sensuous movies remains a pillar for films enriched with local flavor. Some find a way to secure box office reception without regard to markets outside Hong Kong. The warmly received movie *Vulgaria* (低俗喜劇, 2012) earned HK$30 million at the box office and was ranked second in locally produced movies in 2012. However, its foul language and negative depiction of some characters and practices on the mainland ruled out the possibility of entering the coproduction market. A soft pornography movie, *Due West: Our Sex Journey* (一路向西, 2012), which was rated category three for nudity and sexual content, also gain local popularity. A movie with such a rating had no possibility of entering the mainland under China's restrictions of censorship. Nevertheless, it drew attention for its mild success of local theatrical revenues of HK$19 million.

Encouraging signs were observed when *Love in the Puff* (春嬌與志明, 2012) scored satisfactory results both in Hong Kong and on the mainland. In addition to its admirable revenues—HK$28 million and RMB 71 million, respectively—the film's story and style of conventional Hong Kong cinema found an audience across the border with little hindrance from cultural difference. The film is actually a converted sequel of *Love in a Puff*, a romance with a Hong Kong sensibility flavored with foul language. Although minor characters and plotlines related to the mainland were inserted for marketing concerns, *Love in the Puff* preserves strong Hong Kong localism, which is apparently also appealing to mainland audiences. A drama about an aging family servant and her young master, *A Simple Life* (桃姐, 2011), directed by Ann Hui, is another example. The story unfolds with nostalgic compassion for human relationships in Hong Kong's heritage and traditional values. It won critics' acclaim and a series of awards and grossed close to HK$28 million in Hong Kong and RMB 70 million on the mainland. Viewers on the mainland did not reject these movies merely because they focus on characters or plots with a background in Hong Kong, with the treatment of mainland ingredients sidelined.

FILMMAKERS ENGAGING IN COPRODUCTIONS: INTEGRATION OF CINEMATIC STYLE AND THE QUEST FOR CULTURAL HETEROGENEITY

Some filmmakers from Hong Kong find their path to the mainland market to be long and challenging, and their efforts to engage different cultures meet with mixed results. Producer and director Peter Ho-Sun

Chan has been engaging the film market on the mainland since 2005, when his musical *Perhaps Love* (如果．愛, 2005) earned gross revenue of nearly RMB 30 million. *Perhaps Love* was Chan's first coproduction released in China. His latest work, *American Dreams in China* (中國合伙人, 2013), hit RMB 534 million in revenue in the summer of 2013. During a decade working on coproductions on the mainland, Chan has experienced the trend away from big-budget period action-adventure movies rich in visual spectacle toward comedies on contemporary life with moderate production costs. Successful examples in 2012 include *Lost in Thailand* (人再囧途之泰囧). The odd-couple comedy became a top box office hit on the mainland with a gross of RMB 1.26 billion.[38] Another example is *Finding Mr. Right* (北京遇上西雅圖), a romantic comedy set in Seattle in the United States, which holds a box office record for the genre in China with a gross of RMB 518 million.

Chan expressed relief when *American Dreams in China* secured a rather satisfactory box office income. He says the genre allowed him to fulfill his filmmaking potential, while his previous works, mostly action adventures, were undertaken based on marketing objectives.[39] He produced or directed a series of swordsman and kung fu period action movies in the years he cultivated the film market on the mainland. These movies, including *The Warlords* (投名狀, 2007), *Bodyguards and Assassins* (十月圍城, 2009), *Dragon* (武俠, 2011), and *The Flying Guillotines* (血滴子, 2012), were coproductions released both on the mainland and in Hong Kong. Most of them were big-budget action adventures loaded with spectacular battles or expertly choreographed martial-arts fighting scenes.

The pressures of competition and censorship on coproduction filmmakers are well illustrated in *The Warlords*. With production costs exceeding RMB 300 million (nearly US$40 million), *The Warlords* was unable to gain expected profits of RMB 300 million, despite being the first domestic movie to gross over RMB 200 million on the mainland. *Assembly* (集結號, 2007), a war movie by renowned director Feng Xiaogang, hit the screens one week after *The Warlords* and dwarfed it at the box office. It earned a satisfactory gross of nearly HK$30 million in Hong Kong. Like most domestic films from China, coproduced or not, *The Warlords* did not gain much from overseas or non-Chinese-speaking markets. Though the movie swept awards at film festivals in Taiwan and Hong Kong, the fact that it remained unprofitable through its theatrical release put considerable pressure on Chan, who acted as both producer and director.

Controversial issues with the content of coproduction films are also observed in *The Warlords*. Audiences complained about obvious differences between the theatrical version on the mainland and the one released in Hong Kong on DVD. Characters and plots were altered by deleting certain scenes or redubbing dialogue. Audience members who made comparisons said that omissions in the mainland version weakened the integrity of the film's historical context and main characters.[40] Such alterations are usually made at the request of censors. But due to the lack of transparency, no explanation for the alterations was made public. Neither the distributors nor the filmmakers were willing to comment openly about the deletions.

Chan enjoyed reputable results when he launched his film career in Hong Kong in the 1990s. A series of dramas and romantic comedies, including *Alan and Eric: Between Hello and Goodbye* (雙城故事, 1991), *He Ain't Heavy, He's My Father* (風塵三俠, 1993), and *Tom, Dick and Hairy* (新難兄難弟, 1993), marked his delicate sensibility toward relationships between genders and among different generations. His celebrated *Comrades: Almost a Love Story* (甜蜜蜜, 1996) touches on romance amid the hardships of new immigrants from the mainland to Hong Kong. The film earned sound box office revenues and many awards from film festivals. Chan said in an interview that he had to film big-budget blockbusters to avoid being "jobless," because those films were what the mainland market demanded, while the human dramas he adored had to be left aside.[41] After years of genre restraints and marketing pressure, Chan returned to his preferred cinematic style with *American Dreams in China*. His experience as a coproduction filmmaker from Hong Kong represents an uneasy path. It also indicates that the emerging market in China has undergone ten years of transformation through controlled marketization and, to a certain degree, progress toward heterogeneity in form and content.

GLOBALIZING A REGIONAL MARKET: COPRODUCTION FILMS IN CHINA AND INPUT FROM HONG KONG CINEMA

The growth of the film industry in China in practice is a result of the government's changing policy, with goals to achieve economic and cultural leverage regionally and globally. Promoting coproduction films between the mainland and Hong Kong, and with other countries, especially the United States, serves as a window of interaction to uplift the Chinese film industry. Coproduction has integrated the filmmaking resources of Greater China, mostly from Hong Kong, into a regional market with gravity centered on the mainland. Another possible consequence of the development of the Chinese film market is to facilitate the export of Chinese films into overseas markets for more global economic gain as well as more cultural influence from China.

Coproduction filmmaking policy extends beyond Hong Kong. Foreign film companies are eager to market their films in China as coproductions because there is no quota restriction, whereas imported foreign films are limited by a quota system. In addition, such an arrangement is more lucrative than an import arrangement. The production company of an imported foreign film can take 25 percent maximum from revenue made in China, while a coproduction film company can take up to 43 percent of the income, according to regulations. Some foreign film companies attempt to release films under the coproduction category in order to achieve a higher profit share, but the films fall short in meeting the requirements.

To preserve a Chinese presence in investment, personnel, and content, the Chinese authorities are persistent in their requirements. Such requirements mandate one-third of funds from Chinese investors, participation of Chinese actors in key roles, and scenes filmed in China. Senior officials of SARFT call for strict vetting before granting coproduction status, since some "coproduction films" fail to meet the requirements.[42] Controversy reportedly involved some big-budget blockbusters coproduced by investors from China and the United States in 2012. These included *Ironman 3* (Disney) and *Cloud Atlas* (Warner Brothers).

A notable recent tendency is the increasing dominance of financing from Chinese investors, which occurs in coproductions with Hong Kong filmmakers. Some films involve producing initiatives from the mainland side, and the market orientation is principally toward the mainland audience. The majority of financing is from mainland companies, and sometimes there is no investment from the Hong Kong side. An example is *Overheard 2* (竊聽風雲2, 2011), which was financed by investors headed by the mainland's Bona Film Group in Beijing. The film employed filmmakers and a cast from Hong Kong and a story and cinematic style with a Hong Kong background. It grossed RMB 220 million on the mainland and HK$24 million in Hong Kong, a rather satisfactory result in both markets. Bona, the sole production company of the film, announced in June 2013 that it would make another sequel, *Overheard 3*, with the same cast, producer, and directors from Hong Kong. This signals that filmmaking heritage in Hong Kong could be further embedded into the vast mainland market facilitated by investment from the mainland. The category "domestic film" may now be extended to include movies from Hong Kong, underlining further integration of the film industries on the mainland and in Hong Kong.

COPRODUCTION FILMS, HONG KONG CINEMA, AND THE TRANSFORMING NATIONAL FILM INDUSTRY IN CHINA

To meet such challenges, producers and filmmakers of Hong Kong adapt themselves to new conditions if their coproductions are expected to gain approval for distribution and popularity on the mainland. For this reason, the film industry in Hong Kong has cultivated a new creative culture. Filmmakers from Hong Kong tend to find themselves integrating into film industry of a nation where creative liberalism does not necessarily prevail.

The effect of censorship on artistic preference and creative choice is inevitable and should not be ignored. It calls for filmmakers' sensitive eyes to pick out dramatic details relevant to political and ideological issues to avoid results like that experienced with *Shinjuku Incident*. Such problems undoubtedly represent constraints in an environment where filmmakers need to be alert and tiptoeing in their creative process. Another aspect of this issue is how the main melody in filmmaking relates to prevailing government policy. The main melody, art form, and entertainment have been the three pillars of Chinese cinema since the establishment of the nation, as argued by Li.[43] She observes that main melody films have been unable to bring audiences into theaters, despite abundant state funding and extensive policy support. The near collapse of the film industry in the early 1990s prompted the industry's transformation from a planned economy to a market economy. However, the main melody influence remains in Chinese films. Movies like *The Founding of a Republic* (建國大業, 2009) and *The Founding of a Party* (建黨偉業, 2011) were produced as historical epics with sponsorship from the government. Many superstars acted in the films. The themes, plots, and characters were in line with official political and historical interpretations from the government, but the approach has drawn criticism about authenticity.

Another controversy related to the subtle ideological perspective of main melody films was illustrated in *Hero* (英雄, 2002), a film by Zhang Yimou. The story was inspired by the attempted assassination of the ancient Chinese emperor Qin in the third century BC. Domestic and international reviews of the swordsman action movie were mostly favorable. However, a heated debate focused on the plot, in which the assassin aborted his fatal strike on the infamous emperor because he was convinced that Qin's brutality in eliminating his rivals to unify the country would ultimately serve the best interest of the people. J. Hoberman, a film critic of *The Village Voice*, compared *Hero* to *Triumph of the Will*, Leni Riefenstahl's tribute to Nazi leadership in pre–World War II Germany; he questioned the mind-set of glorifying self-sacrifice to uphold ruthless leadership in *Hero*.[44]

Under such conditions, filmmakers from Hong Kong are learning to adapt to the unfamiliar tradition of China's film industry while using international experiences gained from interacting with Hollywood's commercialism and art cinema in Europe. John Woo, Chow Yun Fat, Jackie Chan, Wong Kar-wai, and Yuen Woo-ping are among the filmmakers and stars who have been exposed to Hollywood and have made bold attempts in international markets and achieved valuable experiences. The box office hit and critical success *Internal Affairs* (無間道, 2002), written and directed by Allan Mak and Felix Chong, was acquired by Hollywood for a remake, released in 2006.[45]

Some commercially renowned Hong Kong film directors went global by packaging movies with international flavor, including universal story subjects and scenes set in worldwide locations. Tsui Hark's *Black Mast 2: City of Masks* (黑俠2, 2001) and Ringo Lam's *Undeclared War* (聖戰風雲, 2005) were action adventures shot in English and other languages besides Chinese. A romantic drama in English with a script and cast from Hollywood, *Love Letter* (情書, 1999) was Peter Ho-Sun Chan's early attempt at a breakthrough to the global market.

Hong Kong cinema is searching for a transformed identity with a new localism. In the process it is on China's path of growing regionalism. The integration moves toward restructuring of the national film industry in China, and both developments may not be mutually exclusive. Foreseeing a diversified film market in China with input from a liberalized mentality in filmmaking, Hong Kong cinema could play a significant role by keeping its "native" characteristics. As suggested by Chan, Fung, and Chun, Hong Kong filmmakers, besides using their advantages in production management, international financing, and distribution, should explore new global markets and sources of financing. However, as the film industry in mainland China advances further toward global markets, its reliance on Hong Kong could diminish rapidly.[46]

A national cinema of China as a single entity has been unrealistic for decades due to political and ideological segregation. As Yingjin Zhang suggests, a direct application of a nation-state paradigm to Chinese cinema is also considered problematic.[47] Building a national cinema in China requires a unified media industry with adequate diversity, native localism, and a film market truly open to the public. Taiwan and Hong Kong have been rather alienated from the mainland for political, historical, and cultural reasons since the 1950s. However, well-received romantic dramas and swordsman movies in Mandarin produced by the Shaw Brothers Studio in the 1960s and 1970s somehow linked the audiences in Hong Kong and Taiwan together, despite different dialects and native cultures.

You Are the Apple of My Eye (那些年，我們一起追的女孩, 2011) presented an outstanding example of a Taiwanese movie that gained a warm reception in all areas of Greater China. The nostalgic teen romantic comedy, budgeted at US$1.7 million, scored US$13.7 million in Taiwan. Its box office income was HK$62 million (US$7.9 million) in Hong Kong and RMB 75.8 million (US$2.3 million) on the mainland, and it became the all-time top-grossing Taiwanese movie in both areas. These episodes in film history show that the cultural boundary across the Taiwan Strait is not unbridgeable, as socioeconomic differences have diminished in recent years. It is worth discussing whether in the near future, a national film industry in China, spearheaded by coproduction films, could further lead to a comprehensive and heterogeneous market covering the full extent of Greater China.

The limitations of strict censorship and business competition in China might haunt filmmakers from Hong Kong, the mainland, or elsewhere, while the thriving region provides them with a vast market. With constraints of the political and ideological sense, the extent of marketization and heterogeneous development of a national cinema integrating films from Hong Kong and Taiwan will depend on how far growth can be sustained and nourished. Nevertheless, filmmakers in Hong Kong have to find their way amid such hardships. Meanwhile, the effort to preserve preferences of local audiences in Hong Kong will continue as long as the heritage of a liberal creative culture lingers.

Impacts from difficult environments and political-economic restraints are nothing new to the development of cinema from a historical point of view. Despite setbacks or the repression of creativity, committed filmmakers might find inspiration from such constraints and enrich the cinematic styles they create. As Bordwell and Thompson put it:

> To fully appreciate the films we watch now, we need to be aware that their makers are struggling with the same problems and decisions that appear at every moment of

film history. Technology, tastes, and received traditions offer both opportunities and constraints—sometimes opportunities within constraints.[48]

Note: Special thanks go to John Chong (莊澄), Joe Cheung (張同祖), and Chi Lee (李志毅) for kindly sharing their knowledge and views with me as I wrote this article.

NOTES

1. "Mainland and Hong Kong Closer Economic Partnership Arrangement (CEPA), June 29, 2003. Annex 4: sector 2, Communications Services, part D, Audiovisual Services," Trade and Industry Department, The Government of the Hong Kong Special Administration Region, http://www.tid.gov.hk/english/cepa/index.html.

2. SARFT merged with the General Administration of Press and Publication in March 2013. The restructured office is named State Administration of Press Publication, Radio, Film and Television.

3. Xiaoqing, Zhao (趙小青). "香港電影淡出，中國電影崛起 (Hong Kong Cinema Fades Out Amid The Rise of China's Films)," 香港電影10年："融合與發展 -- 香港回歸十週年電影研討會"論文集，主編：張建勇，張文燕. (北京：中國電影出版社, 2007), 46.

4. "enbase:上半年合拍片市場票房收入達14.2億 (Coproduction Film Grossed 1.42 Billion Yuan In The First Half of 2010)," EntGroup Inc., September 19, 2010, http://www.entgroup.cn/views/a/7970.shtml.

5. Graeme Turner, *Film as Social Practice* (New York: Routledge, 2006), 184.

6. Movies produced in mainland China with themes or content designed to promote political or ideological doctrine; known as *zhu xuan lü dianying* (主旋律電影).

7. Yingchi Chu, "The Consumption of Cinema in Contemporary China," in *Media in China: Consumption, Content and Crisis*, ed. Stephanie Hemelryk Donald, Michael Keane, and Yin Hong (London: RoutledgeCurzon, 2002), 45.

8. Xiaoli Li, *Fu Hua Ying Xiang De Beihou* (Beijing: Communication University of China Press, 2007).

9. EntGroup Inc., *China Film Industry Report 2009–2010 (Compact Version)*, (Beijing: EntGroup Inc., 2010), 15.

10. "*Theatrical Statistics 2012,*" Motion Picture Association of America, Inc., last modified April 2, 2014. http://www.mpaa.org/wp-content/uploads/2014/03/2012-Theatrical-Market-Statistics-Report.pdf.

11. Ibid.

12. "China's 2013 Box Office Nears 21.8 Bln Yuan," Xinhuanet, Jan 3, 2014, http://news.xinhuanet.com/english/china/2014-01/03/c_125954676.htm.

13. EntGroup Inc., *China Film Industry Report*, 8.

14. "《中國廣播電影電視發展報告(2013)》(Report of China's Film and Television Development (2013) ," official website of the Central People's Government of the PRC, July 5, 2013, http://big5.gov.cn/gate/big5/www.gov.cn/jrzg/2013-07/05/content_2440784.htm [accessed June 20, 2014].

15. "2011中國簡況：廣播影視概況 (General Situation of Television and Film in China, 2011),"official website of the Central People's Government of PRC, April 10, 2012, http://www.gov.cn/test/2012-04/10/content_2110108.htm [accessed June 20, 2014].

16. "2012年全國電影票房統計: <表1> 票房收入前10名國產影片(2012 Nationwide Box Office Statistics: <Table 1> Top Ten Domestic Films) ," official website of the State Administration of Press, Publication, Radio, Film and Television, of the PRC, January 11, 2013, http://www.sarft.gov.cn/articles/2013/01/11/20130111112329420341.html.

17. MPAA, *Theatrical Statistics*, 5.

18. David Germain, "Ticket Rush: Film Fans Hand Hollywood Record Cash," *Associated Press,* December 21, 2012.

19. Ibid.

20. "影片歷史排行(All Time Box Office)," EntGroup Inc., June 22, 2014, http://data.entgroup.cn/boxoffice/cn.

21. "All Time Box Office 1-100," Box Office Mojo, June 21, 2014, http://www.boxofficemojo.com/alltime/world [accessed June 20, 2014].

22. Both countries agreed to increase the quota to thirty-four per year in 2013.

23. "2012國內電影市場年均增幅領跑世界票房 (2012 Growth of Film Market in China: No. 1 in the World)," People's Daily Online, May 6, 2013, http://media.people.com.cn/n/2013/0506/c120837-21376498.html.

24. Patrick Frater, "China: Hollywood Studios Still Face Headaches," *Variety*, September 10, 2013.

25. Will Martin, Deepak Bhattasali, and Shantong Li. "China's Accession to the WTO: Impact on China." In *East Asia Integrates: A Trade Policy Agenda for Shared Growth,* edited by Kathie Krumm and Homi Kharas. Washington, D.C.: The World Bank and Oxford University Press, 2004. Pp. 3-20.

26. *Summary of U.S.–China Bilateral WTO Agreement* (Washington, DC: White House Office of Public Liaison, 1999).

27. "《中外合作摄制电影片管理规定》(Regulations on Co-Production Film by China and Foreign Countries)," official website of the State Administration of Press, Publication, Radio, Film and Television, of the PRC, July 6, 2004, http://www.sarft.gov.cn/articles/2004/08/10/20070924090517840608.html.

28. Mainland and Hong Kong Closer Economic Partnership Arrangement.

29. The Film Regulation Ordinance (電影管理條例, Order 342), for example, was issued by the State Council on December 25, 2001.

30. Film Regulation Ordinance, section 24.

31. Ibid., sections 26 and 27.

32. Ibid., section 25.

33. Ibid., sections 25(5) and 25(7).

34. Nearly twenty movie sequels adapting the theme or extended stories of *Young and Dangerous* were produced from 1996 to 2013.

35. David Bordwell, and Kristin Thompson, *Film Art: An Introduction* (New York: McGraw-Hill, 2013), 494.

36. Ni, Zhen (倪震). "香港電影向內地發展的策略和反思 (The Strategy and Review of Hong Kong Film's Development in mainland China)," 香港電影10年：" 融合與發展 -- 香港回歸十週年電影研討會" 論文集, 主編：張建勇，張文燕. (北京：中國電影出版社, 2007), 6.

37. Yingjin Zhang, *Chinese National Cinema* (London: Routledge, 2004), 206.

38. "影片歷史排行(All Time Box Office),"EntGroup Inc., June 22, 2014, http://data.entgroup.cn/boxoffice/cn.

39. Vivien Chen, "Laughing Matter: Contemporary and Romantic Comedies Have Struck a Chord with Mainland Audience," *South China Morning Post,* June 2013, 29–32.

40. "南方都市報：原来我们都被陈可辛骗了(Southern Metropolis Daily: We Are Cheated By Peter Chan)," *Sina,* February 27, 2008, http://ent.sina.com.cn/r/m/2008-02-27/03401927890.shtml [accessed June 20, 2014].

41. Chen, "Laughing Matter."

42. "中外合拍片审查将变严 电影局提三点硬性要求 (Film Bureau Raised Three Rigid Requests: Tougher Vetting for Coproduction Films)," Chinanewsnet, August 27, 2012, http://www.chinanews.com/yl/2012/08-27/4136181.shtml.

43. Xiaoli Li. *Fu Hua Ying Xiang De Beihou*, 68.

44. James Hoberman, "Man with No Name Tells a Story of Heroics, Color Coordination," *Village Voice*, August 17, 2004.

45. The story was adapted for the film *The Departed*, directed by Martin Scorsese and written by William Monahan. The film was released by Warner Brothers.

46. M. Joseph Chan, Anthony Y. H. Fung, and Chun Hung Ng. *Policies for the Sustainable Development of the Hong Kong Film Industry* (Hong Kong: Chinese University Press, 2010), 79.

47. Zhang, *Chinese National Cinema*, 5.

48. Bordwell and Thompson, *Film Art,* 498.

Chapter 18

Political Comics and Freedom of Expression: The 2012 Chief Executive Election in Hong Kong and Implications for Mainland China

Sonny Shiu-Hing Lo

While the rise of political comics in Hong Kong can be traced back to the 1980s, when Britain and the People's Republic of China (PRC) negotiated over the future of the territory's sovereignty, since establishment of the Hong Kong Special Administrative Region (HKSAR) on July 1, 1997, Hong Kong has witnessed the proliferation of political comics.[1] Most importantly, political comics represent the persistence of freedom of expression in the HKSAR, a right that is enshrined in the Hong Kong mini-constitution, the Basic Law. On the other hand, the proliferation of political comics since Hong Kong's return to the PRC has been accompanied by the rise of political opposition and critical discourse in the HKSAR. Although these comics do not represent the views of the majority of Hong Kong citizens, their content as conveyed by cartoonists reflects how some members of the public perceive their government.

THE EMERGENCE AND PERSISTENCE OF POLITICAL COMICS IN HONG KONG

Political comics in Hong Kong can be traced back to the 1960s, when Yen E-King (or Yim Yee-king), a regular editorial cartoonist for *Dai Chung Daily* and *Express Daily,* addressed topics such as the Cultural

Revolution, the Vietnam War, the depreciation of the British pound, the American civil rights movement, and conflicts in the Middle East.[2] Yen's comics were well received, especially his works on the late PRC leader Mao Zedong, Mao's wife Jiang Qing, and the late PRC premier Zhou Enlai. After the Sino–British negotiation over Hong Kong's future began in 1982, political comics flourished in the territory. Zunzi, Ma Long, and Local Boy focused on the details of the 1997 handover of Hong Kong and drew satirical cartoons about the PRC's communist politics. The Tiananmen incident in the PRC on June 4, 1989, sparked a plethora of political comics in Hong Kong under British rule. This situation continued until the transfer of sovereignty over Hong Kong from Britain to the PRC on July 1, 1997. Political cartoonists included Larry Feign, Paul Best, Templar, Harry Harrison, and Gavin Coates.[3]

From the beginning of the HKSAR era, under the leadership of Tung Chee-hwa, to the end of the Donald Tsang period on June 30, 2012, political comics were characterized by severe criticism of government policies, officials, and actions.[4] The difficulties of governance have remained the main theme of political cartoonists in Hong Kong. In particular, former chief executive Tung Chee-hwa became a favorite target of political cartoonists, who adopted a critical and cynical attitude toward his policies and administration. Zunzi depicted article 23 of the Basic Law, which stipulates that the HKSAR government should enact laws outlawing treason, sedition, secession, and subversion, as crucifying the Hong Kong people, who had to shoulder the tax burden under the policies of former financial secretary Anthony Leung (see Figure 18.1).

Figure 18.1

Zunzi exercised his political creativity and critiques in other comics, including one showing that while Anthony Leung appealed to the people of Hong Kong for unity in economic hard times, former secretary for security Regina Ip did not consult citizens over legislation on article 23 (Figure 18.2). Figure 18.3 portrays Ip as an official dressed in red (implying her patriotism toward Beijing) and inviting people into a national security box that could be locked up—a comic severely criticizing the ideology and intention behind the article 23 legislation.

Figure 18.2

Figure 18.3

The Tung Chee-hwa administration was a favorite topic of political comics. Figure 18.4 shows Tung's face hit by a cream cake, with Chinese words saying that the people of Hong Kong would no longer tolerate his leadership and that they should take to the streets. Figure 18.5 shows that Tung introduced the Principal Officials Accountability System (POAS), in which political appointees sharing the chief executive's visions were recruited as the secretaries of his government after July 2002, but the POAS was depicted as a huge problem that Tung could not really "swallow" and that trapped him politically. In reality, POAS was not really accountable to the public, as none of the secretaries were elected by citizens; they were accountable to the chief executive. Figure 18.6, a cartoon drawn by Zunzi, describes Tung as a Soviet-communist-style security minister similar to that of Macao; the implication was that the HKSAR was "Macao-nized" in terms of national security policy, pointing to the retrogression of HKSAR development.

Figure 18.4

Figure 18.5

Figure 18.6

After Donald Tsang became chief executive in 2004, political comics persisted. Financial Secretary John Tsang was portrayed as being eager to distribute cash subsidies to citizens to dilute memories of Hong Kong people about the 1989 Tiananmen incident (Figure 18.7). However, the people of Hong Kong remained defiant about the official verdict from Beijing that the student movement in 1989 was counterrevolutionary. On the night of June 4, 2012, 180,000 people, including local Hong Kong citizens and also many mainland visitors and tourists, participated in a candlelight vigil held at Victoria Park.⁵

Figure 18.7

In Figure 18.8, Ma Lung depicts Tsang as having no idea about a road map of political and democratic reform in Hong Kong while fishing on the issue of double direct elections, namely the direct election of the chief executive through universal suffrage and that of the entire Legislative Council. Figure 18.9, a comic drawn by Zunzi, depicts Tsang as putting up a political show of conducting consultations with the public on the issue of political reform—yet he is working under the constraint of the interpretation of the Basic Law by the Standing Committee of the National People's Congress (SCNPC) of the PRC.

Figure 18.8

Figure 18.9

THE 2012 CHIEF EXECUTIVE ELECTION

The 2012 chief executive election can be seen as a case study to show how comics illustrate the freedom of political expression in the HKSAR. Picture 18.10 shows candidate Henry Tang holding a glass of wine in the face of contender Leung Chun-ying (C. Y. Leung)—a derivative work produced by a netizen. Picture 18.11, drawn by famous political cartoonist Cuson Lo, interestingly depicts the fierce competition between the two top contenders, Tang and Leung, while the third prodemocracy candidate, Albert Ho, does not even show his face entirely. Picture 18.12 elaborates on the key issues in the chief executive campaign, including the controversial architectural design competition in the Western Kowloon District, in which Leung was a panel member accused of neglecting to report on his potential conflict of interests; the dinner involving Leung's key supporters, who met a suspected triad member; the controversy surrounding Tang's illegal home basement; and the critical role in the election played by PRC premier Wen Jiaboa, especially as Wen mentioned in mid-March that the chief executive should ideally be politically acceptable to most Hong Kong people. Wen's remarks implied that public opinion and the political popularity of candidates would be decisive in the race—a comment widely viewed as a watershed in Beijing's changing support from Tang to Leung.

Figure 18.10

Figure 18.11

Figure 18.12

Cuson Lo portrays the chief executive election as "a small circle one," a view shared by many prodemocracy Hong Kong people (Figure 18.13). But he also sees politics as "delightful" in that the race showed how Leung defeated all other potential candidates (Figure 18.14), including Jasper Tsang Yok-sing of the pro-Beijing Democratic Alliance for the Betterment and Progress of Hong Kong (DAB), who at one time toyed with, but eventually shelved, the idea of participating in the election when Tang was enmeshed in scandals such as extramarital affairs and the illegal basement structure at his home.

Figure 18.13

Figure 18.14

Many other derivative works produced by netizens describe the electoral campaign creatively. In Figure 18.15, a netizen uses late PRC premier Zhao Ziyang's appeal to the students on Tiananmen Square in 1989 as a background but replaces the premier's face with that of Tang. This derivative work interestingly portrays Tang as exposing the hard-line attitude of Leung in the July 2003 Executive Council's discussion on how protestors outside the Legislative Council should be tackled. The work exaggerates remarks made by Tang to portray Leung as a hard-line leader who would sacrifice civil liberties of the Hong Kong people. Picture 18.16 replaces Mao's face with Leung's to imply that Leung is a die-hard pro-Beijing Hong Kong leader.

Figure 18.15

Figure 18.16

The most prominent aspect of the 2012 chief executive election was how the scandals surrounding Tang himself stimulated considerable creative, critical, and cynical comics, as well as derivative works. Some netizens even portrayed the defeat of Tang in a harsh manner (Figure 18.17). Tang's extramarital affairs, his penchant for wine, and his illegal basement structure are combined into a comic (Figure 18.18), while the wife of Tang is consistently described as a victim ruthlessly abandoned by Tang himself. He blamed his wife for initiating the illegal basement work, but the mass media and members of the public saw Tang as morally irresponsible, using his wife as a shield to evade political and personal responsibilities. One citizen commented on Figure 18.18 by saying that "Tang, in order to rescue himself from extramarital affairs, had to build illegal structures at his home." The cynical way some netizens portrayed Tang was testimony that many Hong Kong people expected their political leader to be morally honest and to have personal integrity.

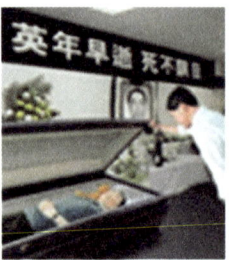

Figure 18.17

Figure 18.18

A large number of political comics and derivative works were critical of Tang's extramarital affairs and the way he handled the illegal basement structure by putting the blame on his wife. Many comics writers and netizens made fun of Henry Tang in a variety of ways, ranging from ridiculing him as a criminal in having illegal home structures to showing him to be a dishonest husband without the boldness to shoulder personal responsibility for making mistakes. Figure 18.19 even portrays Tang as a person who deserves to be put inside a pig cage, a sign that he should be penalized like ancient Chinese women who committed adultery and were put into cages as punishment by villagers and relatives (Figure 18.20). Figure 18.21 portrays Tang as purely a playboy.

Figure 18.19

Figure 18.20

Figure 18.21

As shown in pictures 18.22 to 18.25, netizens were critical of the politics of Hong Kong, including the relatively undemocratic nature of the chief executive election (Figure 18.22), the heavily pro-Beijing political orientation of C. Y. Leung (Figure 19.23), and the deep concern about whether Leung would adopt an authoritarian approach comparable to that of dictators like Adolf Hitler (Figure 18.24). Although the comparisons between Leung and Hitler were exaggerated, such derivative and creative works did highlight the worries of some Hong Kong people about the political impacts of a victory of C. Y. Leung.

Figure 18.22

Figure 18.23

Figure 18.24

The victory of C. Y. Leung in Hong Kong's chief executive election generally raised the concern of many Hong Kong people about the role of the PRC representative office, the Liaison Office, which according to reports interfered with the chief executive election by lobbying election committee members to vote for Leung (Figure 18.25). Finally, some Hong

Kongers were concerned about whether Leung would reintroduce article 23 legislation during his term of office (Figure 18.26).

Figure 18.25

Figure 18.26

From a broader perspective, the rise and persistence of political comics in Hong Kong represent the birth of political symbolism in the opposition movement. Since the massive demonstration of half a million people on July 1, 2003, against the Tung Chee-hwa government, political opposition in the HKSAR has gained considerable strength and growth—a result evident in the September 2012 Legislative Council direct elections, in which the democratic left (those advocating both democracy and welfare) was victorious. The democratic left in the HKSAR, including the League of Social Democrats and People Power, is made up of many young people who are determined to oppose the C. Y. Leung government. Political opposition took an innovative form in the HKSAR when legislators belonging to the democratic left threw toy bananas at the chief executive inside the legislature, employed filibustering tactics to delay the passage of government bills, confronted the police on the streets and in town halls whenever the chief executive and his principal officials went to meet members of the public, and used a toy wolf named Lufsig at Ikea stores to target at C. Y. Leung in a public forum in December 2013.[6]

Clearly, political opposition in the HKSAR takes many forms, including not only comics but also the use of soft toys to criticize the chief executive.

THE INFLUENCE OF HONG KONG'S POLITICAL COMICS ON GUANGZHOU'S POLITICAL CARTOONISTS

According to a Hong Kong interview of two mainland Guangzhou political critics—namely political commentator Chen Yang and political cartoonist Kang Biu—Hong Kong freedom of expression does have an invisible and neglected impact on the mainland, especially in blogs, where citizens and cartoonists utilize political satires to implicitly criticize mainland leaders and officials.[7] While Chen's column in *Southern Metropolitan Daily* criticizes the local government in various aspects of public maladministration, such as the use of paints to repair potholes on the road, Kang has expressed his admiration of Zunzi in capturing the development of Hong Kong politics and has constantly drawn small heads of political figures to criticize "brainless" mainland officials who fail to tackle governance problems.[8] Chen makes an insightful observation that while the people of Hong Kong were interested in materialistic pursuits in the 1980s, they have become far more political than ever before since the 1990s. Conversely, the mainland Chinese, according to Chen, were highly political in the 1960s and 1970s due to the Maoist emphasis on political redness and mobilization, but they have become far more material in their interests since the 1990s. Chen observes a convergence between some Hong Kong people and mainland Chinese in Guangzhou, where they are constantly using political satires and comics to monitor and criticize the performance of local governments—a view shared by Kang. If the observations made by Chen and Kang are accurate, the emergence and proliferation of political comics and cartoonists in Hong Kong have a hidden impact on the mainland, particularly in Guangzhou, where many Cantonese have easy access to Hong Kong's political news, satires, and commentaries through a variety of sources, including blogs, websites, books, and magazines, as well as newspapers that they can buy during visits to Hong Kong. In short, Hong Kong's hidden political impacts on the mainland cannot be underestimated.

THE EMERGENCE OF POLITICALLY CORRECT COMICS IN THE PRC

With the 2013 selection of Xi Jinping, the top PRC president and general secretary of the Chinese Communist Party (CCP), mainland China has witnessed the emergence of some politically correct comics. One website is concerned about how political leaders in the PRC have been recruited, focusing on the Xi Jinping case and his use of forty years to climb up the political ladder from the bottom to the top.[9] Another website shows all the comics concerning anticorruption, a campaign led by Xi to legitimize CCP rule and to purge the corrupt cadres and officials.[10] The website has been shown by the Ministry of Supervision's Central Discipline Inspection Committee (CDIC), focusing on the peasants' drawing of comics against corruption of government officials. The theme of anticorruption is divided into three parts: principles, self-discipline, and new styles of operation. In Qiu County at Hebei Province, 3,000 peasants became comics writers out of a population of 260,000. Each comics writer could earn RMB 5,000 to RMB 6,000 on top of his or her annual income of around RMB 20,000.[11] After reading the comics, some citizens praised the anticorruption efforts. One teacher said, "The comics are like a sharp sword entering the bottom of the heart," implying that corrupt officials and cadres constitute the target and that comics have become an "arts weapon" in the domestic politics of the PRC.[12]

CONCLUDING OBSERVATIONS

Although political comics and derivative works in the HKSAR are not new, their proliferation and persistence since July 1, 1997, and their variety as well as diversity during the 2012 chief executive election can be interpreted as a sign that political cartoonists, comics writers, and netizens are fully exercising their freedom of political expression. Political comics and derivative works demonstrate how Hong Kong people utilize their freedom of expression to create and re-create political satires and to provide critiques of existing political events and systems. Hong Kong netizens and cartoonists have used the existing political space to criticize Hong Kong politics, to criticize Hong Kong's relations with Beijing, and to reinforce the public expectation that political leaders should be morally upright, personally honest, and politically acceptable to ordinary citizens. The fact that Henry Tang's extramarital affairs and illegal basement structure were fully exploited by political cartoonists and netizens shows the sentiment of many Hong Kong people toward the moral conduct of their political leaders. This public opinion also reflects the reality that chief executive candidates in 2012 were elected by only twelve hundred members of the election committee rather than by all Hong Kongers. Because of the legitimacy problem of the elected chief executive, whose mandate came from a narrow membership of the election committee, the people of Hong

Kong who did not have the right to vote for their chief executive had high expectations about the moral and political conduct of the top political leader dealing with the central government in Beijing. Although Beijing perceives the way in which the Hong Kong chief executive was elected in 2012 as conferring upon Leung procedural legitimacy, many Hong Kong people clearly believe that Leung had legitimacy deficits, partly because of the "small circle election" and partly because of his questionable integrity during the hotly contested election campaign.Furthermore, the critical, cynical comics and derivative works demonstrate how the political opposition in the HKSAR perceives Hong Kong's politics in general and the political relations between Beijing and the HKSAR in particular. The highly politicized contents of cartoons, comics, and derivative works in the HKSAR deserve to be observed continuously because they illustrate the persistence of the existing political space, freedom of expression, the extent of political opposition, the development of public opinions, and the moral expectations regarding political leaders in the minds of many Hong Kong people. Finally, although the mainland's political atmosphere is a far cry from Hong Kong's relatively pluralistic and tolerant society, political comics in Hong Kong have hidden impacts on a few mainland political commentators and cartoonists in Guangzhou. Above all, the emergence of political comics has to serve the interests of the CCP, especially its overriding principles of social harmony and political correctness.In any case, the persistent growth of Hong Kong's political comics deserves our further attention and study, for it represents not simply the continuation of freedom of expression in the special administrative region but also a political tradition that has far-reaching repercussions on emergent practices of political comics in China and political satires in Guangzhou in particular. The proliferation of politically incorrect comics in Hong Kong, as opposed to the rise of politically correct comics in the PRC, fully demonstrates the huge contrast between freedom of expression in the HKSAR and the persistence of limited political space in the PRC.

NOTES

1. The author expresses his deepest gratitude to comics writers Zunzi, Ma Lung, and Cuson Lo, whose comics are shown in this chapter, for their kind permission to reprint their works.

2. Wendy Siuyi Wong, *Hong Kong Comics: A History of Manhua* (New York: Princeton Architectural Press, 2002), 80. Also see Wendy Siuyi Wong, "Manhua: The Evolution of Hong Kong Cartoons and Comics," *Journal of Popular Culture* 35, no. 4 (Spring 2002): 25–47.

3. Christine Loh, "Foreword," in Carine Lai Man-yin, *The Rise of Hong Kong Politics: The View through Political Cartoons 1984–2005* (Hong Kong: Civic Exchange, 2006), 29.

4. For details, see Benson Wong, "Analyzing the Ungovernability of Tung-Chee-hwa through Political Cartoons," in *Journal of Local Discourse 2010: New Class Struggle in Hong Kong*, ed. Local Discourse Editorial Committee and SynergyNet (Taipei: Azoth Books, 2011), 163–86.

5. *Apple Daily*, June 5, 2012, S1.

6. Tanna Chong, "Toy wolf with a rude name becomes a must-have for Hong Kong protestors," *South China Morning Post*, December 10, 2013.

7. "Crossing the Mainland," Hong Kong Cable TV Channel, September 9, 2012.

8. Ibid.

9. See "How are leaders trained and recruited?" http://www.youtube.com/watch?v=36k1ogtfRVc (accessed December 10, 2013).

10. See http://v.mos.gov.cn/qiuxian/index.shtml (accessed December 10, 2013).

11. *Ming Pao*, November 8, 2013, A20.

12. *Sing Tao Daily*, November 8, 2013, A36; *The Sun*, November 8, 2013, A46; *Ming Pao*, November 8, 2013, A20.

'All images copyright belong to the artists' although consent from three comic artists had been sought, works by the unknown netizens could not be found.

Conclusion

In conclusion, the development of creative industries in Greater China follows three major models: (1) the minimal role of the Hong Kong government regarding creative potential explored and realized by both individuals and groups; (2) partnerships between the government, groups, individuals, and the market in the case of Taiwan; and (3) the dominant role of the state in the PRC, where the creativity of groups and individuals has to conform with the principles of economic modernization, social harmony, and political correctness. These three models will likely shape the continuous development, transformations, and consolidation of creative industries in the coming decades.

In the case of Hong Kong, the minimal role of the government in the past has slightly changed to include more government-led initiatives. Apart from increased government intervention, corporate planning and community-based art groups persist, so that creative industries blossom in different ways. The abundant talent of groups and individuals flourishes assertively in Hong Kong. Yet without stronger policy directions, Hong Kong's creative industries will not be able to sustain their own momentum. Hong Kong surely needs better discussions, consultation, and formulation of assertive government policies in the development of cultural and creative industries for the sake of enhancing competitiveness.

The partnership between government, individuals, and groups facilitates the growth of various creative industries. This book shows that successful policy transfer is necessary through adapting Western models to local cultural context and market needs. This policy transfer can be applied to the amalgamation of Western educational models and Chinese pedagogy in arts and culture, incorporating urban planning design with cultural characteristics and renaissances in various cities in Greater China. Partnership has become the hallmark of effective governance, business management, and cultural rejuvenation. Creative and cultural industries absolutely demand a partnership model between the government, individuals, and groups.

The dominant role of the PRC government in emphasizing its soft power has contributed immensely to the mushrooming of creative and cultural industries in various cities. The impacts on economic growth are tremendous. However, a country or a city must not lag behind in the process of nurturing creative talents who can produce a special "made in China" branding. Building on the strong technical training foundation in the PRC, the government's development initiatives embrace education, research, and entrepreneurship—with all directed toward the ultimate objectives of stimulating creativity, imaginations, and expression. Indeed, the creativity, imaginations, and expressions of individuals and groups are also shaped by varying contextual circumstances, as the cases of mainland China, Taiwan, Hong Kong, and Macao in this book have demonstrated.

The three models of governance in developing creative and cultural industries unveil the abundant possibilities of developmental pathways, which can be transferred and adapted to different contextual and market needs. All chapters in this book corroborate that these contextual and market circumstances embrace freedom of speech and expression; the nurturing of creative talents; the triangular partnership between the government, individuals, and groups; the advancement of technology; and the policy directions of the government. Undoubtedly, these ingredients are essential in the consolidation, development, and breakthrough of creative and cultural industries.

Index

A

accessibility 204
aesthetics 121, 284, 289
artistic leadership 208

B

bilateral agreements 133, 137, 138
box office 42, 155, 330, 331, 332, 333, 334, 335, 338, 340, 341, 342, 343, 346, 347
branding 36, 39, 68, 70, 76, 77, 102, 203, 204, 205, 207, 210, 301, 374
brush stroke 189, 194, 195, 200

C

calligraphy 19, 20, 38, 118, 121, 129, 183, 184, 185, 186, 187, 188, 189, 190, 191, 192, 193, 194, 195, 196, 197, 198, 199, 200, 201, 228, 233
career development 304, 305, 307
censorship 42, 330, 332, 337, 338, 341, 342, 345, 347
CEPA 17, 153, 155, 157, 158, 252, 329, 331, 336, 337, 349
change agent 218, 224, 225, 227
chief executive election 358, 360, 362, 364, 366, 368
Chinese music 20, 21, 221, 222, 224, 225, 226, 228, 230, 231, 233, 234, 311, 321
cinematic representation 284, 290
cloning 52, 54
cluster 56, 57, 61, 78, 102
clustering 39, 57, 58, 109, 204, 205
commercialism 329, 346
communicator 225
community music 40, 248, 307
community orchestra 219, 231, 235, 236
community theater 248, 249
consumption 50, 51, 56, 57, 167, 168, 287, 295
conventional curricula 316
coproduction film 336, 338, 344
cosmopolitanism 41, 281, 282, 283, 284, 285, 286, 287, 288, 289, 290, 294, 295
couplet 183
creative communities 50, 52, 57, 58
cultural entrepreneurship 15, 20, 39, 203, 208, 211, 218, 219, 220, 222, 231, 236
cultural goods 51, 52, 56, 61
cultural heritage 24, 26, 31, 37, 38, 86, 111, 112, 133, 134, 135, 136, 137, 138, 139, 140, 156, 162, 164, 167, 168, 169, 173, 174, 175, 176, 177, 318
cultural policy 24, 41, 54, 58, 72, 147, 156, 162
cultural property 137, 140
cultural studies 121, 286, 316

D

death penalty 133, 134, 135, 141
derivative works 361, 362, 363, 368, 369
development project 97
differentiation 53, 54
disciplines 23, 24, 118, 124, 208, 303, 319, 324
district 85, 87, 89, 90, 92, 93, 105, 106, 110, 112, 153, 156, 157, 158, 219, 222, 249

E

empirical framework 40, 284, 288, 294, 295, 296
enacted curriculum 41, 317, 322, 323, 324
entrepreneur 48, 207, 217, 218, 225, 300
entrepreneurship educators 303
ethnic folk music 319

F

field theory 284, 285, 290, 294
film industry 42, 152, 153, 155, 158, 329, 330, 331, 332, 333, 343, 345, 346, 347
filmmaker 343
filmmaking 26, 34, 42, 330, 335, 339, 340, 342, 343, 344, 345, 346
flâneur 40, 41, 282, 283, 284, 285, 286, 287, 288, 289, 290, 291, 292, 293, 294, 295, 296
fluid character 40, 283, 288, 294, 295
freedom of expression 27, 40, 42, 353, 367, 368, 369

G

gaze 283, 284, 288, 289

global policy 71, 79

H

Hong Kong cinema 40, 329, 330, 339, 340, 341, 346

I

identity 19, 20, 29, 38, 39, 40, 72, 151, 162, 183, 184, 185, 189, 196, 197, 199, 200, 205, 206, 243, 248, 250, 282, 284, 287, 290, 295, 346
ideograms 200
impact 42, 51, 60, 79, 133, 158, 177, 187, 218, 223, 224, 232, 234, 236, 309, 324, 367
incubation 69, 78
indigenous insider 41, 316, 321, 322, 323, 324
indigenous outsider 316, 319, 323, 324
innovation 17, 47, 49, 50, 52, 53, 57, 58, 61, 69, 73, 78, 90, 119, 151, 152, 155, 216, 223, 228, 229, 250, 323
inscription 184, 186, 190, 194
intended curriculum 317

J

journalism 22

L

lanes 96, 108
liberalism 329, 345
literati painting 118, 119, 128
localities 67, 68, 69, 71, 79
localization 70

M

media 21, 26, 30, 31, 35, 36, 39, 40, 49, 50, 51, 52, 55, 57,

58, 60, 62, 69, 74, 116, 163, 203, 204, 205, 206, 207, 209, 210, 221, 222, 229, 233, 243, 249, 270, 275, 277, 296, 305, 308, 347, 350, 359, 360, 362, 363, 366
metropolitan 288
museum education 162, 167
music curriculum 21, 174, 315, 316, 318, 320, 323, 324
music education 19, 20, 41, 42, 315, 316, 317, 322, 324
music-teacher education 316, 323, 324
music textbooks 317

N

national standards 101
netizen 358, 361
networking 204, 210

P

pedestrianization 36, 85, 97, 103, 105, 107, 111
phono-semantic compounds 200
pillar industry 36, 48, 51
plaza 96, 108
pluralistic societies 324
policy assemblage 65, 70, 71, 74, 76, 79
policy transfer 36, 65, 66, 67, 70, 71, 79, 373
political cartoonist 358, 367
political opposition 353, 366, 367, 369
portfolio 41, 299, 300, 301, 302, 303, 305, 306, 309
preservation 24, 34, 37, 111, 118, 135, 162, 177
producer 55, 89, 155, 284, 342, 344
professionalism 27, 304
proliferation 29, 86, 89, 184, 353, 367, 368, 369
proximity 39, 55, 58, 204, 205, 206, 211
public opinion 359, 369
puppetry 249

R

relics 37, 88, 133, 134, 135, 136, 137, 139, 141, 184, 191, 192, 193, 195
relocation 110
revitalization 36, 85, 87, 89, 94, 95, 99, 104, 105, 111, 184
risk-taker 217, 218, 224, 225, 227

S

seal script 190
small circle election 369
SME 78, 81
social media 26, 39, 203, 206, 207, 210, 221, 222, 233, 305, 308
social network 52, 207, 213
soft power 36, 49, 51, 52, 56, 57, 60, 61, 184, 374
stone carvings 193, 194, 196
sustainability 28, 78, 90, 216, 223, 225, 232, 235, 236
symbolism 184, 185, 190, 191, 199, 366

T

tablet 183, 188, 190, 192, 194, 195
television 31, 34, 48, 56, 68, 223, 229, 231, 246
tourism 24, 35, 49, 77, 99, 101, 102, 103, 104, 154, 167, 168, 173

U

uncertainty 50, 67, 207
upgrading 49, 50, 111, 137

V
venture 207, 217, 230, 302, 333

W
world music 316, 321, 322, 323